MICROBIAL SYNTHESIS OF CHALCOGENIDE NANOPARTICLES

Joyabrata Mal

Joint PhD degree in Environmental Technology

Docteur de l'Université Paris-Est
Spécialité : Science et Technique de l'Environnement

Dottore di Ricerca in Tecnologie Ambientali

UNESCO-IHE
Institute for Water Education

Degree of Doctor in Environmental Technology

Tesi di Dottorato – Thèse – PhD thesis

Joyabrata MAL

Microbial Synthesis of Chalcogenide Nanoparticles

Defended on November 18th, 2016

In front of the PhD committee

Prof. Geoff Gadd	Reviewer
Prof. Davide Zannoni	Reviewer
Dr. Paul Mason	Reviewer
Prof. Piet Lens	Promotor
Dr. Hab. Eric van Hullebusch	Co-Promotor
Dr. Giovani Esposito	Co-Promotor
Prof. Fritz Holzwarth	Examineur

European Commission
ERASMUS
MUNDUS

Erasmus Joint doctorate programme in Environmental Technology for Contaminated Solids, Soils and Sediments (ETeCoS³)

Thesis committee:

Thesis Promotor

Prof. dr. ir Piet N.L. Lens
Professor of Biotechnology
UNESCO-IHE, Delft, The Netherlands

Thesis Co-Promotors

Dr. Hab. Eric D. van Hullebusch,
Hab. Associate Professor in Biogeochemistry
University of Paris-Est, Marne-la-Vallée, France

Dr. Giovanni Esposito,
Assistant Professor of Sanitary and Environmental Engineering
University of Cassino and Southern Lazio, Cassino, Italy

Dr. Y. V. Nancharaiah (Scientific Officer G)
Bhabha Atomic Research Centre
Department of Atomic Energy, Government of India
Kalpakkam, Tamil Nadu, INDIA

Other Members

Prof. Geoff Gadd
College of Life Sciences,
University of Dundee,
Dundee, Scotland, UK

Prof. Davide Zannoni
Dipartimento di Farmacia e Biotecnologie
Università di Bologna,
Bologna, Italy

Dr. Paul Mason
Department of Geosciences
Utrecht University,
Utrecht, The Netherlands

This research was conducted under the auspices of the Erasmus Mundus Joint Doctorate Environmental Technologies for Contaminated Solids, Soils, and Sediments (ETeCoS3) and the Graduate School for Socio-Economic and Natural Sciences of the Environment (SENSE).

Published by:
CRC Press/Balkema
Schipholweg 107C, 2316 XC, Leiden, the Netherlands
Pub.NL@taylorandfrancis.com
www.crcpress.com – www.taylorandfrancis.com
ISBN 978-1-138-60042-3

Imagination is more important than knowledge.
Albert Einstein

Table of Contents

Acknowledgement

Firstly, I would like to thank Erasmus Mundus ETeCoS[3] program (FPA no. 2010-0009) for providing the financial support to carry out this thesis. I would also like to thank UNESCO - IHE, University of Paris-Est, University of Cassino and Southern Lazio, and Université de Limoges for hosting me.

I would like to express my gratitude to Prof. Piet Lens (UNESCO-IHE, The Netherlands), promotor of the thesis, for his constant supervision of the progress of my research and encouragement. His scientific suggestions and constructive comments on the work was critical to achieve success.

A special thanks to Dr. Eric van Hullebusch (University of Paris-Est, France), co-promoter of the thesis, for his support, encouragement and insightful comments in improving the quality of the thesis. I would also like to thank Dr. Giovanni Esposito (University of Cassino and the Southern Lazio, Italy) for his constant support with the administrative affairs related to the PhD program. And many thanks to Dr. Eldon to be always there in need and for your constant support.

I can not be more thankful to Dr. Y. V. Nancharaiah (Venkata), Bhabha Atomic Research Centre, Kalpakkam, India, not only for being my supervisor during the past three years, but also for his friendship. Thanks a lot Venkata for all the help you provided, your comments, your ideas, your scientific and moral support has been instrumental in improving my views on science and the quality of the thesis. I look up to you and the work you do.

I am grateful to Dr. Isabelle Bourven, Dr. Stéphane Simon and Dr. Gilles Guibaud for welcoming me to Limoges and for all their guidance and support during the Short-term scientific mission (COST-ECOST-STSM-ES1302-010216-071163) at Groupement de Recherche Eau Sol Environnement, Université de Limoges, France. Furthermore, I would like to acknowledge with much appreciation the support from the COST action ES1302 for this STSM. I would also like to thank my collaborators Dr. Wouter Veneman, Dr. Willie Peijnenburg and Dr. Martina Vijver (Leiden University). It has been a pleasure knowing you and working with you.

A very special thanks to Prof. Geoff Gadd (University of Dundee), Prof. Davide Zannoni (University of Bologna) and Dr. Paul Mason (Utrecht University), jury members, for their critical comments and and suggestions which led me thinking more deeply about my research and had provided me with a new perspective of my thesis. Moreover, it has also given me thoughts on how to improve this work further. I would like to express my gratitude towards Prof. Fritz Holzwarth for chairing my PhD defense session.

I would also like to express my gratitude to Dr. Chloé Fourdrain (University of Paris-est) and Dr. Santanu Bera (BARC, India) for helping with the Raman, XPS and XRD analysis. My gratitude also goes to IHE- lab staff, Fred, Berend, Ferdi, and Lyzette. Thank you guys for all your help in the lab. Thanks to Frank for his help in reactors set up and the pumps. Thanks to Peter, it was really fun to work with you in lab and will never forget our "serious" discussion on footbaal. We will keep supporting Arsenal together.

This section is incomplete without acknowledging my colleagues and now friends Nirakar, Shrutika, Feishu, Susma, Samayita, Tejaswini, Lea, Niranjan, Angelica, Mullele, Mohaned, Mohan, Motasem, Manivannan, Neeraj and Paolo. I would like to thank Chiarra and Arda, it had been great fun to share office and lab bench with you. It has been a pleasure knowing you and working with you. A warm hearted thanks to Joseph (my PhD buddy) for introducing me to IHE and Delft in the beginning, I know I could always count on you. It has been a lot of fun and, of course, many important "non scientific" discussions with you all. Also, a very special thanks to Rohan, Suthee and Erika to introduce me to the selenium world and for their encourgament. Many thanks to the colleagues Clément, Douglas, Taam, Andreina in Paris and Anne, Nathalie and Asmaa in Limoges.

Finally, I would like to express my deepest gratitude to my family for their blessings and love. Thanks to my parents for supporting me to chase my dreams despite the distance and time. Without your constant support and sacrifice, this would not have been possible. This section will not be complete without thanking Bhawna for her love, encouragement, and understanding. You have been a great support for me and thanks for living with me during my worst phases. You have always been supportive and also to someone that I can look upto and hope to keep knowing you for the rest of my life. Thanks a lot Debasish uncle, Rahul, Sonia and bubai for encouraging me every day and most of all to the Almighty God who made everything possible.

Summary

Recent years have seen a growing interest in the application of chalcogenide nanoparticles (NPs) (e.g. Se, Te) in various industrial sectors including energy, steels, glass and petroleum refining. The fluorescent metal chalcogenide (e.g. CdSe, CdTe) NPs are used in solar cells, optoelectronic sensors and also in the field of biology and medicine for imaging or sensing including biolabelling. Moreover, due to the high toxicity of chalcogen oxyanions (i.e., selenite, selenate, tellurite and tellurate), their release in the environment is of great concern. Thus, emphasize was given in this thesis on the development of a novel microbial synthesis process of chalcogenide NPs by combining biological treatment of Se/Te-containing wastewaters with biorecovery of Se/Te in the form of Se/Te chalcogenides NPs.

A special focus was given to study the effect of heavy metal (e.g. Cd, Zn and Pb) co-contaminants on selenite bioreduction by anaerobic granular sludge. Anaerobic granular sludge capable of reducing selenite to selenide in the presence of Cd was enriched for the microbial synthesis of CdSe NPs. It was evident that when Cd is present along with selenite, it either forms a Se-Cd complex by adsorption onto biogenic Se(0) nanoparticles after Se-oxyanion bioreduction or it reacts with aqueous selenide (HSe⁻) to form CdSe. The absorption and fluorescence spectra of the aqueous phase confirm the presence of CdSe NPs. Raman spectroscopy and X-ray photoelectron spectroscopy (XPS) analysis support this finding. The formation of an alloyed layer of CdS_xSe_{1-x} at the interface between the CdSe core and CdS shell in the sludge was also observed. Detailed studies on the extracellular polymeric substances (EPS) reveal that the protein and polysaccharide-like content increased in the EPS extracted from enriched sludge while humic-like substances decreased. Size exclusion chromatography (SEC) of EPS further reveals a distinct fingerprint for proteins and humic-like substances, with increase in high molecular weight protein-like and the appearance of new peaks for humic-like substances in the EPS after the enrichment.

An upflow anaerobic granular sludge bed (UASB) reactor was used for the first time for continuous removal of tellurite from synthetic wastewater and the recovery of Te as biogenic Te(0). Energy-dispersive X-ray spectroscopy (EDS), X-ray diffraction (XRD) and Raman spectroscopic analysis of biomass confirmed the deposition of Te(0) in the biomass. It was evident that the majority of the Te(0) was trapped predominantly in the EPS surrounding the biomass, which can be easily separated by centrifugation.

Sommaria

Gli ultimi anni hanno visto un crescente interesse per l'applicazione di nanoparticelle calcogenuri (NP) (ad esempio Se, Te) in diversi settori industriali, tra cui l'energia, acciaio, vetro e raffinazione del petrolio. Il calcogenuro metallico fluorescenti (ad esempio CdSe, CdTe) NP sono utilizzati nelle celle solari, sensori optoelettronici e anche nel campo della biologia e della medicina per l'imaging o rilevamento compreso biolabelling. Inoltre, a causa della elevata tossicità di ossianioni calcogeno (vale a dire, selenite, selenate, telluriti e tellurate), il loro rilascio nell'ambiente è di grande preoccupazione. Così, enfatizzano è stato dato in questa tesi sullo sviluppo di un nuovo processo di sintesi microbica di NP calcogenuri combinando il trattamento biologico di SE / Te-contenenti acque reflue con biorecovery di Se / Te in forma di selenio / Te calcogenuri NP.

Una particolare attenzione è stata data per studiare l'effetto di metalli pesanti (ad esempio Cd, Zn e Pd) co-contaminanti su bioreduction selenite dal fango granulare anaerobica. fango granulare anaerobica grado di ridurre selenite di seleniuro in presenza di Cd è stato arricchito per la sintesi microbica dei CdSe NP. Era evidente che, quando Cd è presente con selenite, esso sia forma un complesso Se-Cd per adsorbimento su biogenico Se (0) nanoparticelle dopo Se-oxyanion bioreduction o reagisce con seleniuro acquosa (HSe-) per formare CdSe. Gli spettri di assorbimento e fluorescenza della fase acquosa confermano la presenza di CdSe NP. Spettroscopia Raman e X-ray spettroscopia fotoelettronica (XPS) sostegno dell'analisi questo risultato. La formazione di uno strato legato di CdSxSe1-x all'interfaccia tra il nucleo e CdSe CdS guscio in stato inoltre osservato fanghi. Studi dettagliati sulle sostanze polimeriche extracellulari (EPS) rivelano che il contenuto di proteine e polisaccaridi come sono aumentati negli EPS estratte dal fango arricchito mentre sostanze umiche-simile sono diminuiti. esclusione Dimensione cromatografia (SEC) di EPS ulteriori rivela un'impronta distinta per proteine e sostanze umiche simili, con aumento alta proteine come il peso molecolare e la comparsa di nuovi picchi delle sostanze umiche, come nei EPS dopo l'arricchimento.

Un anaerobica letto fango granulare (UASB) reattore di flusso ascendente è stato usato per la prima volta per la rimozione continua di tellurito da acque reflue sintetica e il recupero di Te come biogenica Te (0). Energia dispersiva a raggi X spettroscopia (EDS), diffrazione di raggi X (XRD) e Raman analisi spettroscopica di biomassa confermato la deposizione di Te (0) nella biomassa. Era evidente che la maggior parte del Te (0) è stato intrappolato prevalentemente EPS circondano la biomassa, che può essere facilmente separato mediante centrifugazione.

Samenvatting

De laatste jaren is de belangstelling voor de toepassing van chalcogenide nanodeeltjes (NP) toegenomen (bv. Se, Te) in verschillende industriële sectoren, zoals voor energie, staal, glas en aardolieraffinage. De fluorescerende metaalchalcogenide (bv. CdSe, CdTe) NP worden gebruikt in zonnecellen, opto-elektronische sensoren en in de biologie en geneeskunde voor beeldvorming of detectie, inclusief biolabelling. Bovendien, door de hoge toxiciteit van chalcogen anionen (d.w.z. seleniet, selenaat, telluriet en telluraat), is hun introductie in het milieu van groot belang. Aldus werd in dit onderzoek de ontwikkeling van een nieuwe microbiële synthese van chalcogenide NP onderzocht, door een combinatie van biologische behandeling van Se/Te bevattend afvalwater en biorecovery van Se/Te in de vorm van Se/Te chalcogenide NP.

Speciale aandacht werd besteed aan het effect van zware metalen (bv. Cd, Zn en Pd) co-contaminanten op seleniet bioreductie door anaëroob korrelslib. Anaeroob korrelslib dat seleniet reduceert tot selenide in de aanwezigheid van Cd werd opgehoopt voor de microbiële synthese van CdSe NP. Het was duidelijk dat wanneer Cd samen met seleniet aanwezig is, het of een Se-Cd complex vormt door adsorptie aan biogene Se (0) nanodeeltjes na Se-oxyanion bioreductie of het reageert met waterig selenide (HSe-) om CdSe te vormen. De absorptie en fluorescentie spectra van de waterfase bevestigen de aanwezigheid van CdSe NP. Raman spectroscopie en röntgen foto-elektron spectroscopie (XPS) analyse ondersteunen deze bevinding. Vorming van een gelegeerde laag CdSxSe1-x bij de interface tussen de CdSe kern en CdS schil werd ook waargenomen in het slib. Uitvoerige studies naar de extracellulaire polymere substanties (EPS) tonen aan dat de eiwit en polysacharide inhoud toenemen in de EPS geëxtraheerd van het verrijkte slib, terwijl de humus-achtige verbindingen verminderden. Deeltjesgrootte-uitsluitingschromatografie (SEC) van EPS toonde voorts een duidelijke vingerafdruk voor eiwitten en humus-achtige verbindingen, met een toename van hoog molecuulgewicht eiwit-achtige stoffen en het verschijnen van nieuwe pieken bij de humus-achtige substanties in het EPS na de verrijking.

Een opstroom anaërobe granulaire slib bed (UASB) reactor werd voor het eerst voor de continue verwijdering van telluriet uit synthetische afvalwater en voor de terugwinning van tellurium als biogene Te(0) gebruikt. Energiedispersieve röntgenspectroscopie (EDS), röntgendiffractie (XRD) en Raman spectroscopische analyse van de biomassa uit deze UASB reactor bevestigde de depositie van Te(0) in de biomassa. Het was duidelijk dat de meeste

Te(0) overwegend in EPS rondom de biomassa aanwezig was, die gemakkelijk door centrifugatie kon worden afgescheiden.

Résumé

Ces dernières années, un intérêt grandissant a été porté à l'application des nanoparticules (NPs) de chalcogène (par exemple, Se, Te) dans divers secteurs industriels tel que l'énergie, l'acier, le verre et le raffinage du pétrole. Les NPs fluorescentes de chalcogéne métallique sont utilisées dans les cellules photovoltaïques, les capteurs optoélectriques et dans le domaine de la biologie et de la médecine pour l'imagerie et les techniques de détection utilisant le marquage biologique. De plus, de part la forte toxicité des oxyanions chalcogènes (sélénite, sélénate, tellurite et tellurate), leur déversement dans l'environnement est préoccupant. Ainsi, cette thèse s'est concentrée sur le développement d'un nouveau procédé de synthèse microbienne de NPs de chalcogène, qui combine un traitement biologique d'eaux usées contenant du Se/Te avec la récupération biologique du Se/Te sous la forme de NPs de Se/Te.

Il a notamment été étudié l'effet des métaux lourds (par exemple, Cd, Zn et Pd) comme co-contaminant sur la réduction biologique par des boues granulaires anaérobies. Les boues granulaires anaérobies capables de réduire les sélénites en sélénides en présence de Cd ont été enrichies pour la synthèse microbienne de NPs de CdSe. Il a été clairement observé que lorsque le Cd est présent en même temps que les sélénites, il se forme soit un complexe Se-Cd par adsorption à la surface des NPs biogénique de Se(0) après la bioréduction des oxyanions de Se, soit il y a une réaction avec les séléniures aqueux (HSe$^-$) menant à la formation de CdSe. Les spectres d'absorption et de fluorescence de la phase aqueuse confirment la présence de NPs de CdSe dans la phase aqueuse. Les analyses de spectroscopie Raman et de spectroscopie photoélectronique par rayons X sont en accord avec ces observations. La formation d'une couche d'alliage de CdS_xSe_{1-x} à l'interface entre le noyau des CdSe et la surface extérieure des CdS dans les boues a aussi été observée. Des études détaillées des substances polymériques extracellulaires (SPE) ont révélé que le contenu en protéine et en molécule de type polysaccharide a augmenté dans les SPE extraites des boues enrichies, alors que le contenu en substances de type humique a diminué. La chromatographie d'exclusion de taille (CET) des SPE a aussi révélé des caractéristiques distinctes pour les protéines et les substances de type humiques, avec une augmentation de la quantité de molécules avec des poids moléculaire élevés pour les protéines et l'apparition de nouveaux pics pour les substances de types humiques dans les SPE après l'enrichissement.

Un réacteur à lit de boue granulaire anaérobie à flux ascendant a été utilisé pour la première fois pour l'élimination continue des tellurites contenues dans des eaux usées synthétiques et pour la récupération du Te come Te(0) biogénique. Les analyses de la biomasse par spectroscopie à rayons X à dispersion d'énergie, par diffraction des rayons X et par spectroscopie Raman ont confirmé la déposition de Te(0) à la surface de la biomasse. Il a été clairement observé que la majorité des Te(0) a été piégée de manière prédominante dans les SPE entourant la biomasse, qui peuvent être séparées facilement par centrifugation.

CHAPTER 1

General introduction

Chapter 1

1.1. Background

Selenium (Se) and tellurium (Te) belong to the chalcogen group (periodic table group 16) and share many common characteristics. Both are metalloids and exist in four oxidation states, i.e. +VI, +IV, 0, and −II in nature (Mal et al., 2016a; Zannoni et al., 2008). Often the Se and Te oxyanion contamination occurs through anthropogenic activities such as mining, refinery industries and coal combustion and electronic industries (Belzile & Chen, 2016; Mal et al., 2016a). Oxyanions (selenate, selenite and tellurate, tellurite) of both elements are soluble and mobile, thus bioavailable and toxic (Nancharaiah & Lens, 2015; Turner et al., 2012). Compared to that, elemental Se and Te are non-soluble and less toxic. Bioreduction of Se/Te oxyanions to elemental Se(0) and elemental Te(0), respectively, can be achieved using both aerobic and anaerobic microorganisms and microbial reduction of chalcogen oxyanions is gaining considerable interest for bioremediation and treatment of Se/Te wastewaters due to their relatively low cost, non-toxic and environmental friendly approach (Mal et al., 2016b; Nancharaiah & Lens, 2015; Nancharaiah et al., 2016; Turner et al., 2012).

Unlike bulk materials, nanoparticles show peculiar physical, chemical, electronic and biological properties due to their high surface to volume ratio, large surface energy, and spatial confinement and reduced imperfections. Recent years have seen a growing interest in the application of chalcogenide nanoparticles (Ch NPs) (e.g. Se, Te) due to their physicochemical properties in various industrial sectors including energy, steels, glass and petroleum refining (Mal et al., 2016b). The metalloid chalcogens are also used in the preparation of fluorescent metal chalcogenides (MeCh), like cadmium selenide (CdSe) and cadmium telluride (CdTe) NPs, which are particularly suited for their use in solar cells and optoelectronic sensors. MeCh NPs are widely used in the field of biology and medicine for imaging, sensing including fluorescent biolabelling and cancer detection (Mal et al., 2016b).

1.2. Problem description

The synthesis of Ch NPs with different size and shape is a key goal but remains a challenge in nanotechnology. Several physical and chemical methods exist for synthesis of highly stable Ch NPs with defined properties but remain a major concern due to the use of toxic chemicals on the surface of nanoparticles along with non-polar solvents in the synthesis procedure (Mal et al., 2016b). Also most of these methods use energy intensive procedures with highly toxic precursors and reagents. Hence, in order to meet the requirements and exponentially growing demand, there is a need to develop green methods which are environmental friendly by using renewable materials instead of toxic and hazardous chemicals.

Moreover, both Se and Te are found in relatively low abundances in the earth's crust. Selenium is present in the earth's crust at an estimated amount of 0.05 - 0.5 mg Se. kg^{-1}, while the relative abundance of Te is even lower than that of gold, platinum and other rare earth elements and is estimated to be only 1 - 5 µg. kg^{-1} (Belzile & Chen, 2016; Tan et al., 2016). Thus, development of new technologies is essential for the recovery of these two elements from waste streams and its end-use applications to ensure its future availability. Microbial reduction is a proven technology for converting soluble chalcogen oxyanions to insoluble forms to remove the pollution from the water phase (Nancharaiah et al., 2016; Turner et al., 2012). Thus emphasize should be given on developing microbial technologies for Ch NPs synthesis by combining remediation of Se/Te-containing wastewaters and consequently recovery of Se/Te in the form of Se/Te chalcogenide nanoparticles is gaining considerable interest (Mal et al., 2016b).

Particularly metal chalcogenide (e.g. metal selenide) quantum dots (QDs) have attracted considerable attention due to their quantum confinement and size-dependent photoemission characteristics (Mal et al., 2016b). In order to meet the requirements and exponentially growing demand, there is a need to develop green methods which are environmental friendly by using renewable materials instead of toxic and hazardous chemicals. However, for microbial synthesis of metal selenide nanoparticles (e.g. CdSe), there is not enough understanding on the effect of heavy metals on selenium bioreduction. Also, no study has so far been carried out on the detailed characterization of the fate and speciation of heavy metals and bioreduced selenium in the presence of selenium reducing microorganisms.

Application of Ch NPs recovery at the industrial scale requires a continuous cost-effective technology. Reduction of selenate was demonstrated in a UASB reactor for the treatment of selenium containing wastewater (Dessì et al., 2016; Lenz et al., 2008). But until now, tellurium recovery using anaerobic granular sludge by continuous reactors has not yet been reported. Since extracellular polymeric substances (EPS) directly contact and interact with metals in natural environments, they are of vital importance not only for protecting the interior microbial cells, but also for geochemical cycling and remediation of metals in natural environments factors (Li & Yu, 2014; Raj et al., 2016). Special attention should be given to EPS-metal interactions enabling removal and recovery of Ch NPs. Also it is unclear whether EPS of biofilms associated with biogenic Ch NPs influence the fate, bioavailability and toxicity of the Ch NPs in natural environments.

1.3. Research objectives

The overall objective of this thesis is the microbial synthesis of chalcogenide nanoparticles (e.g. Se(0), CdSe, ZnSe, PbSe and Te(0)) by combining the bioremediation of Se/Te-containing wastewater with biorecovery of Se/Te in the form of Se/Te chalcogenide nanoparticles.

The specific research objectives are:

1) To utilize selenium reducing microbial communities for removing selenite from wastewater and to recover Se in the form of elemental Se and CdSe NPs:
   a) Biological removal of selenate and ammonium by activated sludge in a sequencing batch reactor
   b) To investigate microbial reduction of selenite in the presence of heavy metals and analyze the fate of the bioreduced selenium and the heavy metals
   c) Enrichment of microorganisms capable of reducing selenite to selenide in the presence of Cd for the synthesis of CdSe NPs
   d) To develop a microbial synthetic protocol for the production of metal selenide nanoparticles.

2) To investigate the difference in EPS fingerprints extracted from enriched anaerobic granular sludge in the presence of Cd and Se (IV):
   a) EPS extraction and detailed characterization to have in-depth understanding on the composition and molecular size fingerprint of EPS
   b) To differentiate protein and humic-like substances fingerprints based on the apparent molecular weight (aMW) distribution in EPS samples by coupling size exclusion chromatography (SEC) coupled to UV and fluorescence detectors
   c) To elucidate EPS - metal(loid) interactions i.e. Cd and Se/CdSe

1. To investigate tellurite (Te(IV)) reduction by a mixed microbial community, i.e. anaerobic granular sludge and recovery of biogenic Te(0):

    a) To evaluate the feasibility of a laboratory-scale upflow anaerobic granular sludge bed (UASB) reactor inoculated with anaerobic granular sludge for continuous bioreduction of Te(IV)

    b) Recovery and characterization of biogenic Te(0)

2. To investigate the toxicity of biogenic nano-Se (nano-Se[b]) formed by anaerobic granular sludge biofilms on zebrafish embryos in comparison with selenite and chemogenic nano-Se (nano-Se[c]):

    a) Synthesis of biogenic nano-Se (nano-Se[b]) and bovine serum albumin (BSA) stabilized nano-Se[c]

    b) To study the fate of both nano-Se in terms of shape, size, aggregation and dissolution kinetics

    c) To investigate the toxicity of selenite, nano-Se[b], and nano-Se[c] on zebrafish embryos

    d) To determine the effect of particle size on nano-Se toxicity

1.4. Structure of the thesis

This dissertation is comprised by nine chapters. An overview of the structure of this dissertation is given in Fig. 1.1. The following paragraphs outline the content of the chapters presented in this dissertation.

Chapter 1 presents a general overview of this dissertation, including background, problem description, research objectives and thesis structure.

Chapter 2 provides the literature review about the current status of the biological synthesis and applications of metal chalcogenide nanoparticles. The role of key biological macromolecules in controlled production of the nanomaterials is highlighted. The technological bottlenecks of microbial synthesis of metal chalcogenide limiting widespread implementation are also discussed.

Chapter 3 explores the potential of simultaneous removal of ammonium ($NH_4^+ - N$) and selenate (Se(VI)) using activated sludge in a sequencing batch reactor (SBR).

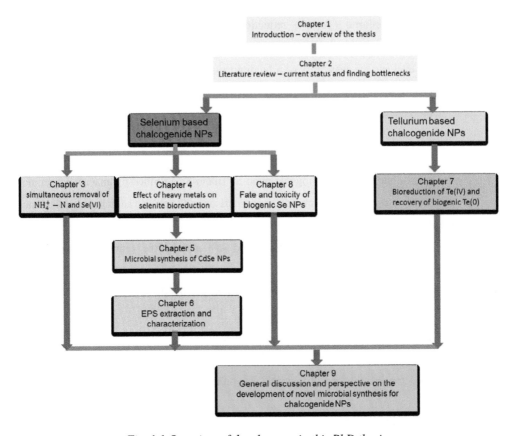

Fig. 1.1 Overview of the chapters in this PhD thesis

Chapter 4 presents the bioreduction of selenite by anaerobic granular sludge in the presence of heavy metals and analyzed the fate of the bioreduced selenium and heavy metals.

Chapter 5 describes the enrichment of a microbial community in anaerobic granular sludge capable of reducing selenite to selenide in the presence of Cd for the synthesis of CdSe NPs.

Chapter 6 investigates the detailed characterization and change in EPS after the enrichment of anaerobic granular sludge in the presence of cadmium (Cd) and selenite for microbial synthesis of CdSe.

Chapter 7 evaluates the feasibility of a laboratory-scale upflow anaerobic granular sludge bed (UASB) reactor inoculated with anaerobic granular sludge for continuous bioreduction of Te(IV) and recovery of biogenic Te(0).

Chapter 8 investigates the toxicity of nano-Se[b] formed by anaerobic granular sludge biofilms on zebrafish embryos in comparison with selenite and chemogenic nano-Se (nano-Se[c]).

Chapter 9 summarizes and draws conclusions on knowledge gained from this dissertation. It also gives recommendations for future research.

References

Belzile, N., Chen, Y.-W. 2016. Tellurium in the environment: A critical review focused on natural waters, soils, sediments and airborne particles. *Appl Geochem.*, **63**, 83-92.

Dessì, P., Jain, R., Singh, S., Seder-Colomina, M., van Hullebusch, E.D., Rene, E.R., Ahammad, S.Z., Lens, P.N.L. 2016. Effect of temperature on selenium removal from wastewater by uasb reactors. *Water Res.*, **94**, 146-154.

Lenz, M., van Hullebusch, E.D., Hommes, G., Corvini, P.F., Lens, P.N.L. 2008. Selenate removal in methanogenic and sulfate-reducing upflow anaerobic sludge bed reactors. *Water Res.*, **42**, 2184-2194.

Li, W.W., Yu, H.Q. 2014. Insight into the roles of microbial extracellular polymer substances in metal biosorption. *Bioresour Technol.*, **160**, 15-23.

Mal, J., Nancharaiah, Y.V., van Hullebusch, E.D., Lens, P.N.L. 2016a. Effect of heavy metal co-contaminants on selenite bioreduction by anaerobic granular sludge. *Bioresour Technol.*, **206**, 1-8.

Mal, J., Nancharaiah, Y.V., van Hullebusch, E.D., Lens, P.N.L. 2016b. Metal Chalcogenide quantum dots: biotechnological synthesis and applications. *RSC Adv.*, **6**, 41477-41495.

Nancharaiah, Y.V., Lens, P.N.L. 2015. Ecology and biotechnology of selenium-respiring bacteria. *Microbiol Mol Biol Rev.*, **79**, 61-80.

Nancharaiah, Y.V., Mohan, S.V., Lens, P.N.L. 2016. Biological and bioelectrochemical recovery of critical and scarce Metals. *Trends Biotechnol.*, **34**(2), 137-155.

Raj, R., Dalei, K., Chakraborty, J., Das, S. 2016. Extracellular polymeric substances of a marine bacterium mediated synthesis of CdS nanoparticles for removal of cadmium from aqueous solution. *J Colloid Interface Sci.*, **462**, 166-175.

Tan, L., Nancharaiah, Y.V., van Hullebusch, E.D., Lens, P.N.L. 2016. Selenium: environmental significance, pollution, and biological treatment technologies. *Biotechnol Adv.*, **34**(5), 886-907.

Turner, R.J., Borghese, R., Zannoni, D. 2012. Microbial processing of tellurium as a tool in biotechnology. *Biotechnol Adv.*, **30**(5), 954-963.

Zannoni, D., Borsetti, F., Harrison, J.J., Turner, R.J. 2008. The bacterial response to the chalcogen metalloids Se and Te. *Adv Microb Physiol.*, **53**, 1-12.

CHAPTER 2

Literature review

This chapter has been published in modified form:

Mal, J., Nancharaiah, Y.V., van Hullebusch, E.D., Lens, P.N.L. 2016. Metal Chalcogenide quantum dots: biotechnological synthesis and applications. RSC Adv. 6, 41477-41495

Abstract

Metal chalcogenide (metal sulfide, selenide and telluride) quantum dots (QDs) have attracted considerable attention due to their quantum confinement and size-dependent photoemission characteristics. QDs are one of the earliest products of nanotechnology that were commercialized for tracking macromolecules and imaging cells in life sciences. An array of physical, chemical and biological methods have been developed to synthesize different QDs. Biological production of QDs follow green chemistry principles, thereby use of hazardous chemicals, high temperature, high pressure and production of by-products is either minimized or completely avoided. In the past decade, significant progress has been made wherein a diverse range of living organisms, i.e. viruses, bacteria, fungi, microalgae, plants and animals have been explored for synthesis of all three types of metal chalcogenide QDs. However, better understanding of the biological mechanisms that mediate the synthesis of metal chalcogenides and control the growth of QDs is needed for improving their yield and properties as well as addressing issues that arise during scale-up. In this review, we present the current status of the biological synthesis and applications of metal chalcogenide QDs. Where possible, the role of key biological macromolecules in controlled production of the nanomaterials is highlighted, and also technological bottlenecks limiting widespread implementation are discussed. The future directions for advancing biological metal chalcogenide synthesis are presented.

Keyword: Quantum dots, chalcogenide, metal sulphide, metal selenide, metal telluride, biological synthesis,

2.1. Introduction

Research and development on inorganic nanomaterials such as metal, metal oxide and semiconductor nanoparticles has emerged into a cutting edge multidisciplinary nanotechnology (Narayanan & Sakthivel, 2010; Schröfel et al., 2014). Particularly, semiconductor metal chacogenide quantum dots (QDs) have attracted wide interest for applications from material science to medicine (Azzazy et al., 2007; Valizadeh et al., 2012). Physical and chemical methods exist for synthesis of highly stable metal chalcogenide nanoparticles (MeCh NPs) with defined properties (Biju et al., 2008; Mussa & Valizadeh, 2012). But most of these methods use energy intensive procedures with highly toxic precursors and reagents. On the other hand, microorganisms are known to create a variety of nanomaterials in a most energy efficient manner using non-toxic, inexpensive and renewable reagents. Therefore, biotechnological synthesis of nanomaterials is increasingly considered for the production of nanomaterials in an environmental friendly way according to green chemistry principles (Lloyd et al., 2011; Narayanan & Sakthivel, 2010). This review presents the state of art research on microbial synthesis of metal chalcogenide nanoparticles, the properties of biogenic metal chalcogenides and their applications. The challenges and perspectives of microbial synthesis of metal chalcogenides are also discussed.

2.2. Quantum dots and their properties

QDs are semiconductor nanoparticles made up of elements from groups II-VI or III-V of the periodic table and characterized as inorganic colloids with physical dimensions of 1 to 20 nm, smaller than the bulk-exciton Bohr radius, i.e. the distance in an electron-hole pair (Rogach, 2008). Due to this very small size, the QDs exhibit distinct optical and electronic properties (Alivisatos, 1996; Murray et al., 1993). Since the bandgap energy depends on the particle size, when the particle size is in the nano-scale, optical properties such as fluorescence excitation and emission can be "tuned" by altering the particle size, shape or surface structure (Zhang et al., 2003). QDs have attracted applications as inorganic fluorophores because of bright, "size-tunable" fluorescence with narrow symmetric emission bands, and high photostability (Rzigalinski & Strobl, 2009). The distinct separation between the excitation and emission spectra of the QDs make them better with the detection sensitivity, as the entire emission spectra of QDs can be detected (Jaiswal & Simon, 2004).

11

2.3. Metal Chalcogenide QDs

Chalcogens are the chemical elements of the group 16 of the periodic table such as oxygen, sulfur (S), selenium (Se), tellurium (Te) and polonium (Po). The metalloid chalcogens such as S, Se and Te are used in semiconductors and in the preparation of MeCh. Among the various colloidal semiconductor nanoparticles, MeCh NPs such as CdS, ZnS, CdSe, and CdTe have attracted considerable attention due to their quantum confinement effects and size dependent photoemission characteristics (Joo et al., 2003). The fluorescent MeCh NPs are superior to organic fluorophores in terms of narrow absorption and emission spectra, quantum yield and photostability (Mussa & Valizadeh, 2012). The photooptical and photovoltaic properties of MeCh are particularly suited for their use in solar cells and optoelectronic sensors (Gaponik et al., 2002; Trindade et al., 2001). QDs are widely used in the field of biology and medicine for imaging, sensing (Gao et al., 2010a; Gao et al., 2004; Medintz et al., 2005; Michalet et al., 2005; Park et al., 2011; Sapsford et al., 2006; Smith et al., 2008) and tracking particles or cells (Chang et al., 2008; Maa et al., 2014) including fluorescent biolabelling and cancer detection (Jie et al., 2011).

MeCh NPs are synthesized using various methods such as microwave heating (Li et al., 2014d; Wada et al., 2001), microemulsion (Ohde et al., 2002; Talapin et al., 2002), ultrasonic irradiation (Zhua et al., 2014) and several other chemical methods (Hodlur & Rabinal, 2014; Luoa et al., 2014; Yang et al., 2009). Newer methods are being developed to produce high quality MeCh NPs and to make the synthetic procedures simpler and scalable. In general, most of the current synthetic procedures use hazardous solvents, explosive precursors, and high temperatures (230 - 250°C) (Aguiera-Sigalat et al., 2012; Murray et al., 1993). In order to meet the requirements and exponentially growing demand, there is a need to develop green methods which are environmental friendly by using renewable materials instead of toxic and hazardous chemicals. In this context, synthesis of metal chalcogenide nanoparticles by microorganisms has attracted interest as a viable alternative for the chemical synthesis routes.

2.4. Biological metal chalcogenides

Almost all organisms (viruses, bacteria, fungi, algae and animals) and some biomolecules have been studied for their role as catalysts or nucleating agents for synthesis of MeCh. The vast majority of the studies focused on the use of microorganisms such as bacteria and fungi because of their rapid growth, ease of maintenance and handling (Narayanan & Sakthivel, 2010). In anaerobic conditions, microorganisms oxidize organic compounds by utilizing chalcogen oxyanions (e.g. sulfate or selenate) as terminal electron acceptors in their metabolism. In this process, sulfur, selenium and tellurium oxyanions are reduced to the corresponding chalcogenides sulfide, selenide and telluride, respectively. Microbially produced chalcogenide reacts with the dissolved metal ions (e.g. Zn, Cd, and Pb) to form MeCh NPs (Fig. 2.1).

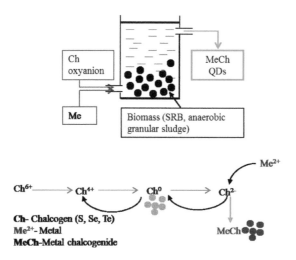

Fig. 2.1 General schematic for biological synthesis of metal chalcogenide QDs

Biological methods for manufacturing nanomaterials are in general considered safe, cost-effective and environment-friendly processes (Ayano et al., 2013; Kumar et al., 2007). The main aim is to develop a protocol to produce nanomaterials that can be incorporated into high-performance products that are less hazardous to human health or to the environment. The biological methods offer advantages such as synthesis at ambient temperatures and pressures, use of renewable materials as electron donors, use of inexpensive microorganisms, production of biocompatible nanomaterials and a possibility to use waste as precursor materials. The disadvantages include poor control on size, shape and crystallinity, scalability and separation of the nanoparticles from the microorganisms are sometimes difficult task. Another major drawback is the slow production rate compared to the chemical synthesis techniques, making the biological production more time consuming (Hosseini et al., 2013). The limitations of biogenic methods can be improved by proper selection of microorganisms, optimizing the reaction conditions such pH, incubation temperature and time, concentration of metal ions, and biomass content (Narayanan & Sakthivel, 2010).

A large numbers of reviews on the chemical and biological synthesis of metal nanoparticles have been published, but information on biogenic metal chalcogenides is scattered and the challenges in the production of biogenic MeCh are not addressed (Hosseini et al., 2013; Jacob et al., 2016). The review on microbial synthesis of chalcogenide by Jacob et al. (2016) is a mini-review and mainly focuses on MeSe QDs. In the currently submitted review, we present for the first time in details the recent developments in biological synthesis of three kinds of MeCh, i.e. MeS, MeSe and MeTe) QDs. This manuscript is also the first comprehensive review on biological synthesis MeTe QDs.

Another important aspect is that the authors have put more focus to distinguish two different mechanisms of biological synthesis of MeCh QDs: i) using hazardous or reactive chemicals and ii) synthesis of MeCh QDs by converting environmentally toxic wastes into MeCh QDs. So, more emphasize was given on the biological synthesis of MeCh QDs by combining the bioremediation approach with synthesis of MeCh QDs, which makes this review distinctly different from other, previously published review papers. The authors critically reviewed almost all significant research in the field and provide a hypothesis of the biological mechanisms for the synthesis of MeCh QDs and to control the growth of QDs. This will certainly help in biological synthesis of "Cd-free" QDs (e.g. InSe) also in near future.

2.5. Sulfur based chalcogenides

Sulfur is one of the most abundant elements on Earth. The sulfur cycle is complex, as it exists in different oxidation states, varying from sulfide (-II) (completely reduced) to sulfate (+VI) (completely oxidized) (Muyzer & Stams, 2008). Sulfate reducing bacteria (SRB) play an important part in the sulfur, carbon and nitrogen cycles. SRB are cosmopolitan in distribution, often found in a variety of environments such as soils, sediments and domestic, industrial and mining wastewaters. They are important members of microbial communities with economic, environmental and biotechnological interest (Chang et al., 2001). SRB reduce sulfate to sulfide, which reacts with divalent metals, such as cadmium and zinc, forming insoluble metal sulfide (MeS) precipitates (Benedetto et al., 2005).

2.5.1. Sulfur oxyanion reduction mechanisms

SRB are anaerobic microorganisms that are widespread in anoxic habitats and use sulfate as a terminal electron acceptor for their growth (Grein et al., 2013). In the dissimilatory sulfate reduction pathway, prior to reduction, sulfate is activated by an ATP sulfurylase, resulting in the formation of adenosine-phosphosulfate (APS) and pyrophosphate. The formation of APS is an endergonic reaction and is driven by hydrolysis of the pyrophosphate by pyrophosphatase to form 2-phosphate (Grein et al., 2013; Muyzer & Stams, 2008). Two ATP molecules are consumed in the activation of sulfate to APS. Subsequently, APS is reduced to sulfite by the APS reductase (AprBA), a heterodimeric iron–sulfur flavoenzyme. The final step of the sulfate reduction pathway is the conversion of sulfite to sulfide, which is catalyzed by dissimilatory sulfite reductase (DsrAB) with the involvement of the small protein DsrC. Some important biological sulfur conversions in the presence of various electron donors are summarized in Table 2.1 (Muyzer & Stams, 2008).

Table 2.1 Energetics of biological sulfate reduction and sulfur disproportionation reactions

Redox reaction	$\Delta G_r'$ (kJ/mol)
$SO_4^{2-} + 4H_2 + H^+ \rightarrow HS^- + 4H_2O$	-151.9
$0.5\ SO_4^{2-} + (C_6H_5O_3)^- \rightarrow 0.5\ HS^- + HCO_3^- + (C_2H_3O_2)^-$	-80.2
$SO_4^{2-} + (C_2H_3O_2)^- \rightarrow HS^- + 2\ HCO_3^-$	-47.6
Conversion and disproportionation	
$4S(0) + 4H_2O \rightarrow 3HS^- + SO_4^{2-} + 5H^+$	-4.6

2.5.2. Biological synthesis of metal sulfides QDs

Biosynthesis of metal sulfides (MeS), an important type of semiconductor nanomaterials, has been one of the most appealing research fields. The higher stability of sulfides in contrast to other chalcogenide compounds, and their wide band gap makes them more suitable for industrial applications including high-temperature operations, high voltage optoelectronic devices, and high efficiency electric energy transformers and generators (Hosseini & Sarvi, 2015). The first demonstration of biological synthesis of quantum semiconductor crystallites of CdS was reported using *Bacillus cereus* (Dameron et al., 1989). Subsequent studies have used several bacteria, fungi, yeast, algae, plants and viruses for the biogenic production of MeS NPs (Table 2.2) (Peltier et al., 2011).

2.5.2.1. MeS QDs synthesis using bacteria

Microbial synthesis of MeS NPs usually requires raw materials of metal and sulfide ions as precursors which can be supplied as soluble salts. In several studies of MeS synthesis, sodium sulfide or hydrogen sulfide was used as the precursor material (Pandian et al., 2011; Prasad & Jha, 2010). For example, *E. coli*, *B. casei* SRKP2 and *Lactobacillus* sp. were incubated with Na$_2$S and CdCl$_2$ to form CdS nanoparticles (Castuma et al., 1995; Pandian et al., 2011; Prasad & Jha, 2010; Sweeney et al., 2004). Since sulfide was used as the source of sulfur, there was no requirement for a reducing agent or SRB. The microorganisms acted as the support for complexation, nucleating centers and templates for metal sulfide seeds and growth of nanoparticles. Metal ions cause stress conditions due to which microorganisms secret stress proteins as a defense tool. These secreted proteins bind with the metal cations, and subsequently with HS$^-$ in solution leading to growth of MeS nuclei and formation of MeS NPs.

The synthesis of CdS NPs strongly depends on the growth phase of the cells. A higher yield of CdS NPs was observed in the stationary phase of the bacterial cultures. The reason could be the enhanced production of proteins and polyphosphate during the stationary phase which plays a role in CdS nanocrystal formation and growth (Castuma et al., 1995; Rao & Kornberg, 1996). An experiment with four different *E.coli* strains (ABLE C, TG1, RI89, and DH10B) for synthesis of CdS showed the effect of genetic differences among bacterial strains on the nucleation of nanocrystals (Sweeney et al., 2004). Only the *E. coli* ABLE C strain was able to produce CdS NPs suggesting that genetic differences among bacterial strains strongly affect the nucleation of nanocrystals. Although, not much information was available, it was reported that synthesis of polyphosphate, increased at stationary phase in *E. coli* ABLE C strain (Sweeney et al., 2004). Polyphosphate is an in vitro nanocrystal capping agent and may possibly act as a nanocrystal templating agent and could be one reason why only the *E. coli* ABLE C strain was able to produce CdS NPs.

Table 2.2 Summary of microorganisms, reaction conditions, and properties of biogenic metal sulfide nanoparticles

Microorganisms	Precursor materials	Temperature (°C)	pH	NPs	Size (nm)	Morphology	Location of synthesis	Ref.
Bacteria								
K. pneumoniae	$Cd(NO_3)_2$, SO_4^{2-}	37	7.6	CdS	200	Sphere	Extra cellular	Holmes et al. (1997)
Thermoanaerobacter sp	$CdCl_2$, $S_2O_3^{2-}$	65	6.9	CdS	10	Hexagonal	Extra cellular	Moon et al. (2014)
Pseudomonas spp.	$CdCl_2$, $S_2O_3^{2-}$	15	7.0	CdS	10-40	Spherical	Cell envelope	Gallardo et al. (2014)
B. casei SRKP2	$CdCl_2$, Na_2S	25	7.0	CdS	10-20	Spherical	Intra cellular	Pandian et al. (2011)
B. amyloliquifaciens KSU 109	$Cd(NO_3)_2$, Na_2S	30	7.2	CdS	3-4	Spherical	-	Singh et al. (2011)
E.coli	$CdCl_2$, Na_2S	25	7.0	CdS	6	-	Intra cellular	Mi et al. (2011)
Lactobacillus sp	$CdCl_2$, H_2S	-	-	CdS	5	Spherical	Extra cellular	Prasad & Jha, (2010)
R. palustris	$CdSO_4$	30	7.0	CdS	8	-	Extra cellular	Bai et al (2009)
E. coli	$CdCl_2$, Na_2S	37	7.2	CdS	2-5	Spherical	Intra cellular	Sweeney et al. (2004)
Bacillus cereus	$CdSO4$	37	-	CdS	30–100	-	Intra cellular	Harikrishnan et al. (2014)
Bacillus subtilis	$CdCl_2$, Na_2S	35	-		2.5–5.5	Spherical	Extra cellular	El-Shanshoury et al. (2012)
Phormidium tenue	$CdCl_2$, Na_2S	-	-	CdS	5	Spherical	Extra cellular	Mubarakali et al. (2012)
Desulforibrio caledoiensis	$Zn(NO_3)_2$, Na_2SO_4	30	7.4	ZnS	30	Spherical	Extra cellular	Qia et al. (2013)
Desulfovibrio sulfuricans	$ZnSO_4$	22	7.2	ZnS	-	-	-	Peltier et al. (2011)
Desulfobacteriaceae sp.	$ZnSO_4$	-	-	ZnS	2-5	Spherical	Intra cellular	Labrenz et al. (2000)
Desulfovibrio	$ZnSO_4$	25	7.0	ZnS	20-30	Spherical	Extra cellular	da Costa et al. (2012)

	Precursor	Temp	pH	QD	Size	Shape	Location	Reference
desulfuricans							cellular	
Thermoanaerobacter sp.	$ZnCl_2$, H_2S	65	7.8	ZnS	2–10	Spherical	Extracellular	Moon et al. (2014)
R. sphaeroides	$ZnSO_4$	30	6.8	ZnS	8	Spherical	Extracellular	Bai et al. (2006)
S. nematodiphila	$ZnSO_4$	35	-	ZnS	80	Spherical	Extracellular	Malarkodi & Annadurai, (2013)
R. sphaeroides	$PbCl_2$	30	7.0	PbS	10.5	Spherical	Extracellular	Bai & Zhang, (2009)
Fungi/yeast								
S. pombe	$CdSO_4$	30	5.6	CdS	1–1.5	Spherical	Intracellular	Williams et al. (2002)
S. cerevisiae	$CdCl_2$, H_2S	-	-	CdS	3.5	Spherical	Extracellular	Prasad & Jha, (2010)
C. glabrata	$CdCl_2$, SO_4^{2-}	30	5.8	CdS	-	-	Intracellular	Krumov et al. (2007)
S. pombe	$CdCl_2$, SO_4^{2-}	30	5.8	CdS	-	-	Intracellular	
P. chrysosporium	$Cd(NO_3)_2$, C_2H_5NS	37	7.0	CdS	2.5	Spherical	Extracellular	Chen et al. (2014)
S. pombe	$CdSO_4$	-	5.6	CdS	2–2.5	-	Intracellular	Kowshik et al. (2002a)
C. versicolor	$Cd(NO_3)_2$, Na_2S	25	5.6	CdS	5–9	Spherical	Extracellular	Sanghi & Verma, (2009)
S. cerevisiae	$ZnSO_4$	25	-	ZnS	30–40	Spherical	Intracellular	Sandana & Rose, (2014)
R. diobovatum	$Pb(NO_3)_2$	30	5.6	PbS	2–5	-	Intracellular	Seshadri et al. (2011)
Torulopsis sp.	$Pb(NO_3)_2$	30	5.6	PbS	4–8	Spherical	Intracellular	Kowshik et al. (2002b)
Virus								
Bacteriophage P22 VLP	$CdCl_2$, C_2H_5NS	30	7.6	CdS	40	Spherical	Intracellular	Zhou et al. (2014)
Algae								

C. reinhardtii	CdCl$_2$, K$_2$SO$_4$	28	3.5	CdS	-	-	Intra cellular	Edwards et al. (2013)
C. merolae	CdCl$_2$, K$_2$SO$_4$	45	3.5	CdS	-	-	Intra cellular	
Plant biomass								
S. lycopersicum	CdSO$_4$	25	5.8	CdS	4-10	Spherical	Intra cellular	Al-Shalabi et al. (2014)

In contrast, when sulfate is used as the sulfur source, the mechanism of MeS NPs formation is different depending on assimilatory or dissimilatory reduction of the dissolved sulfate ions. Several microorganisms mediate sulfur transformations using dissimilatory reduction pathways and generate thus sulfide from mine waters and industrial effluents which contain high concentration of sulfate and heavy metals. These waters can be treated using combinations of bacterial sulfate reduction to generate sulfide, followed by removal of heavy metals as MeS precipitates (Azabou et al., 2007; Gallegos-Garcia et al., 2009; Jong & Parry, 2003; Kaksonen et al., 2003). Sulfate-reducing biofilms and suspensions have been successfully applied for the production of the ZnS in a fluidized-bed reactor of acidic wastewater containing sulfate and zinc (Kaksonen et al., 2003).

Some bacteria follow a different mechanism wherein they use other sulfur source for forming MeS nanocrystals. Bacterial strains such as *Klebsiella pneumonia* (Holmes et al., 1997) and *K. planticola* Cd-1 (Sharma et al., 2000) are able to reduce thiosulfate and form CdS nanoparticles when grown in a medium amended with sodium thiosulfate and Cd(II) ions. *Rhodopseudomonas palustris*, a purple non-sulfur bacterium was used to synthesize CdS (Bai et al., 2009), while *Rhodobacter sphaeroides* was used for production of ZnS (Bai et al., 2006) and PbS (Bai & Zhang, 2009) NPs using cysteine as sulfur source. Cysteine desulfhydrase (C-S-lyase) of phototropic bacteria plays a role in CdS nanocrystal formation. Cysteine rich proteins can produce HS⁻ through the action of C-S-lyase (Wang et al., 2000). The production of CdS nanocrystals was higher during the stationary phase because total C-S-lyase activity was almost double in stationary phase cells as compared to other growth phases (Bai et al., 2009). But overproduction of C-S-lyase can be toxic to cell growth, which makes C-S-lyase production a key parameter in optimizing the sulfide production. In addition, Cd also strongly influences the production of cysteine and activity of the C-S-lyase. Sulfide production and cadmium removal both become adversely affected at higher concentrations of Cd (100 and 125 µM) (Wang et al., 2000). Hence, along with production of the C-S-lyase, the Cd concentration also needs to be optimized for maximum sulfide production and CdS precipitation.

A metal reducing thermophilic bacterium *Thermoanaerobacter* was used for scalable production of ZnS at 65°C (Moon et al., 2014) using thiosulfate as sulfur source. Thermophilic strains confer certain advantages including fast reaction and HS⁻ formation as compared to mesophilic or psychrotolerant microorganisms (Moon et al., 2007). However, the energy incurred to maintain high temperatures should be considered for cost effective production. On the contrary, Gallardo et al. (2014) demonstrated the production of CdS QDs from thiosulfate by using a psychrophilic Antarctic bacterium (*Pseudomonas* spp.) at 15°C. A time-dependent variation of fluorescence color was also observed at 15°C, switching from green to red emission. QDs synthesis at low-temperature provides certain advantages such as a better control on size and polydispersity of NPs based on a kinetic dependent-nucleation process (Meixner et al., 2001). Moreover, if it is possible to biosynthesize QDs at low temperatures with the same yield as compared to high temperature, the energy costs would decrease enormously.

2.5.2.2. MeS QDs synthesis using yeast/fungi

Schizosaccharomyces pombe has been studied extensively for the formation of CdS NPs using sulfate as the sulfur source (Kowshik et al., 2002a; Krumov et al., 2007; Williams et al., 1996a; Williams et al., 1996b). Earlier studies suggested that timing of addition of heavy metals is very important for MeS production. Addition of heavy metals during the early-exponential phase can affect the growth and cellular metabolism and may cause an enhanced efflux of metals from the cell, possibly in the form of unstable MeS NPs. Heavy metal addition during the stationary phase does not result in significant MeS NPs production as the metal uptake and intracellular sulfide production is much less during the stationary phase. It was suggested that heavy metal addition during the mid-exponential growth phase is desired for the formation of stable MeS NPs with a good yield (Williams et al., 1996b).

The formation of biogenic MeS NPs by fungi also depends on factors such as high heavy metal uptake rate, appropriate intracellular heavy metal storage and a large biomass production (Williams et al., 1996b; Williams et al., 2002). The initial glucose concentration and glucose consumption strongly affects the biomass yield. Although an increase in the initial glucose concentration leads to a higher final biomass concentration during the stationary phase, much of the glucose is converted to ethanol leading to negative effects on the cellular metabolism. Interestingly, the excess glucose concentration results in higher specific cadmium uptake rates which lead to higher CdS NPs production (Krumov et al., 2007; Williams et al., 2002). The glucose concentration thus needs to be optimized for minimal glucose repression and ethanol production, but maximal biomass production and metal uptake to achieve a maximum MeS QDs yield. Thus, fed-batch fermentation of yeast is ideal for the synthesis of MeS QDs production as it can minimize glucose repression under a controlled specific growth rate for the cells, enabling high cell densities and product formation (Williams et al., 2002). Williams et al. (2002) reported that the final *S. pombe* biomass concentration (dry weight) of 18.2 g l^{-1} achieved under fed-batch conditions was four times higher than the final biomass concentration in batch experiments.

The formation of metal sulfides in fungal cells generally takes place through a few independent steps. First is the formation of low molecular weight phytochelatin–metal ion complexes as phytochelatin chelate with cytoplasmic cadmium ions. It prevents aggregation of toxic metal ions and their accumulation in specific cell organelles (Chen et al., 2014; Krumov et al., 2007). Then, the chelating compounds are either complexed to form metal-binding particles on the cell wall or transported across the vacuolar membrane via an ATP binding cassette (ABC)-type vacuolar membrane protein, i.e. HMT1. Finally, HS$^-$ is generated by enzymes in the purine biosynthesis pathway and reacts with the phytochelatin-metal complexes to form phytochelatin-metal sulfide complexes or metal sulfide NPs (Kowshik et al., 2002a; Seshadri et al., 2011; Speiser et al., 1992). Other than metal-peptide interactions, at the same time, other carboxylic acid-containing biomolecules such as proteins or polypeptides might also be responsible for capping the nanocrystals via hydrogen bonding and electrostatic interactions (Sanghi & Verma, 2009). Interestingly, *Candida glabrata* has a slightly different particle formation mechanism. *C. glabrata* employs detoxification of heavy metals (e.g. Cu and Zn) by metallothioneins except for Cd and the number of (γ-GluCys) residues in the phytochelatins is different which could be the main reason behind the lower accumulation of CdS NPs (Mehra & Winge, 1991).

One major drawback of fungal synthesis of sulfur based NPs is that the NPs are produced intracellular and the timing of the recovery of intracellular NPs from batch cultures is a very critical step because of cell lysis. If delayed, QDs could be "lost" to the growth medium, making retrieval and purification more difficult (Williams et al., 1996a). In contrast, *Phanerochaete chrysosporium* (Chen et al., 2014) has been successfully exploited for extracellular CdS NPs production, which makes it much easier for harvesting the QDs (Fig. 2.2). HRTEM image (Fig. 2.3) showed that *P. chrysosporium* was able to produce uniform sphere-shaped CdS QDs with an average particle size of 1.96 ± 0.1 nm and the mycelial surface is a superior place for the self-assembly of the CdS nanoparticles. Although reports on the fungal synthesis of nanoparticles are available, focusing more on the biosynthesis's mechanism could enable better control over the biosynthesis process allowing the preparation of high quality nanomaterials.

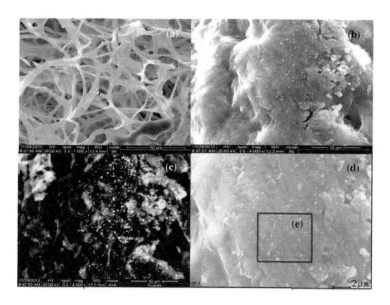

Fig. 2.2 SEM–EDX micrograph of mycelium pellets: (a) native; (b) and (c) treated with Cd^{2+}; (d) expanded image of the nanocrystals in (b); e) showing plenty of nanoparticles were adsorbed on the mycelia surface (Adapted from Chen et al., 2014)

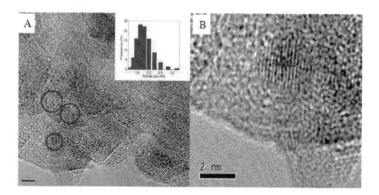

Fig. 2.3 (A) HRTEM images of extracellularly biosynthesized CdS QDs. (Inset) Size distribution of the QDs. (B) Expanded image of QDs. (2 2 0) lattice fringes of denoted area (d_{220} = 2.1A°) (Adapted from Chen et al., 2014)

2.6. Selenium based chalcogenides

Selenium has been referred to as an "essential toxin" due to its requirement as a trace element in living systems and potential toxicity at only slightly higher concentrations (Lenz & Lens, 2009). Selenium is a potential contaminant of concern in natural aquatic environments primarily because of large scale anthropogenic activities. In natural environments, Se exists in four different oxidation states: Se(VI), Se(IV), Se(0) and Se(-II). Microbial activities significantly contribute to the natural selenium cycle and are responsible for Se transformations in different oxic, anoxic and anaerobic environments (Nancharaiah & Lens, 2015b). In fact, microbial reduction of soluble selenium oxyanions to insoluble elemental selenium has emerged as a leading technology for bioremediation and treatment of Se wastewaters (Nancharaiah & Lens, 2015b). Further reduction of Se will lead to the formation of Se(-II) in natural and engineered settings. Under reducing conditions, selenide reacts with the co-existing heavy metal ions and forms insoluble metal selenide precipitates (e.g., PbSe, CdSe, ZnSe, and FeSe) or polysulfideselenide complexes ($-S_nSe_n-$) (Herbel et al., 2003; Weres et al., 1989).

2.6.1. Selenium oxyanion reduction mechanisms

Reduction of Se oxyanions (selenate and selenite) is widespread in natural environments. It is more likely that biotic mechanisms, such as assimilatory and dissimilatory selenium reduction, are responsible for the presence of selenide in the environment (Herbel et al., 2003; Nancharaiah & Lens, 2015b). Dissimilatory reduction and assembly of selenium oxyanions through anaerobic respiration is a two-step process involving the formation of elemental selenium nanoparticles: reduction of selenate to selenite is catalyzed by a trimeric molybdoenzyme, SerABC selenate reductase, located in the periplasmic space. Selenite is then reduced to elemental selenium, mediated by multiple mechanisms that include glutathione and glutaredoxin (Nancharaiah & Lens, 2015b). It appears that elemental selenium nanoparticles are formed via divergent mechanisms and localized both in the cytoplasm and outside the cells. In several studies, elemental selenium nanospheres were observed as the stable end products of microbial reduction (Srivastava & Mukhopadhyay, 2013; Zhang et al., 2011).

In addition, enzymes such as nitrite reductase, sulfite reductase, fumarate reductase, and hydrogenase I catalyze the reduction of selenite to elemental selenium (Li et al., 2014; Nancharaiah & Lens, 2015b; Stolz et al., 2006). For example, *B. selenitireducens* mediated selenite reduction involves energy conservation by lactate oxidation coupled to growth via respiratory reduction of selenate (Switzer et al., 1998). Fe(III) reducers such as *Shewanella oneidensis* (Klonowska et al., 2005) and *Geobacter sulfurreducens* reduce Se (IV) to Se(0) via c-type cytochromes. Unlike *S. oneidensis, Veillonella atypica* produces Se nanospheres from selenite via a hydrogenase coupled reduction, mediated by ferredoxin (Johns, 1951).

Microorganisms conserve energy when selenate is used as the terminal electron acceptor of anaerobic respiration. The Gibbs free energy change for dissimilatory reduction of selenate, selenite and elemental Se(0) is given in Table 2.3 (Baumle et al., 2004; Herbel et al., 2003). Clearly, the reduction of Se(VI), Se(IV) and Se(0) are exergonic reactions favoring energy conservation and growth of microorganisms. Although the reduction of Se(0) yields less potential energy than similar respiratory reduction reactions of Se(VI) or Se(IV), the energy yields increase at alkaline pH and in the presence of Fe(II). However, the disproportionation reactions for Se are clearly unfavorable at either neutral or alkaline pH, as well as in the presence of Fe(II) suggesting that disproportionation does not contribute significantly to Se(-II) formation in natural environments and microbial cultures (Herbel et al., 2003).

Reduction of selenate and selenite has been observed in several microorganisms, but Se(-II) formation was noticed only in a few selenite reducing bacterial cultures. For example, formation of trace amounts of selenide was observed in experiments on selenate reduction using *Salmonella entericaserovar* Heidelberg, *Clostridium pasteurianum, Desulfovibrio desulfuricans* and cell extracts of *Micrococcus lactilyticus* (Steinberg & Oremland, 1990; Woolfolk & Whiteley, 1962). Zehr and Oremland showed that the sulfate-reducing bacterium *D. desulfuricans* and anoxic estuarine sediments reduced trace amounts of Se(VI) to Se(-II) in the presence of sulfate (Zehr & Oremland, 1987). Herbel et al. (2003) reported that *B. selenitireducens*, a Se(VI)-respiring bacterium, produced significant amounts of Se(-II) from Se(0) or Se(VI). Pearce et al. (2009) reported that *G. sulfurreducens* and *V. atypica* are capable of reducing Se(IV) to Se(-II), but *S. oneidensis* can reduce selenite only up to Se(0). While *G. sulfurreducens* exhibited a continuous reduction of Se(IV) to Se(-II), *V. atypica* followed a biphasic reduction reaction. The production of Se(-II) occurred only after complete reduction of Se(IV) to Se(0). Surprisingly, the reduction of Se(0) to Se(-II) is not observed in several other Se(VI) reducing bacteria, i.e. *Sulfurospirillum barnesii, B.*

arseniciselenatis, and *Selenihalanaerobacter shriftii*. It remains unclear why Se(0) reduction to selenide is not observed in these Se(VI)-respiring bacteria. Compared to the soluble selenate or selenite, reduction of insoluble Se(0) is challenging and bacteria may need to employ specific electron transport systems to perform reduction of nanosized Se(0) deposits. Selection of bacteria with particular attributes is thus a vital factor for microbial production of Se(-II) required for metal selenide synthesis.

Table 2.3 Energetics of biological selenium reduction and disproportionation reactions (Herbel et al., 2003)

Redox reaction	$\Delta G_r'$ (kJ/mole e−)
$2SeO_3^{2-} + (C_6H_5O_3)^- + H^+ \rightarrow Se_{Amorp} + (C_2H_3O_2)^- + HCO_3^- + H_2O$	-64.7
$2SeO_4^{2-} + (C_6H_5O_3)^- \rightarrow Se_{Amorp} + (C_2H_3O_2)^- + HCO_3^- + H^+$	-107.4
$2Se_{Amorp} + (C_6H_5O_3)^- + 2H_2O \rightarrow 2HSe^- + (C_2H_3O_2)^- + HCO_3^- + 3H^+$	−11.9
$2Se_{Amorp} + (C_6H_5O_3)^- + 2Fe^{2+} + 2H_2O \rightarrow 2FeSe + (C_2H_3O_2)^- + HCO_3^- + 5H^+$	−30.1
Conversion and disproportionation	
$Se_{Amorp} \rightarrow Se_{black\ hex}$	−3.35 kJ/mole
$4Se_{Amorp} + 4H_2O \rightarrow 3HSe^- + SeO_4^{2-} + 5H^+$	+61.3
$4Se_{Amorp} + 3Fe^{2+} + 4H_2O \rightarrow 3FeSe + SeO_4^{2-} + 8H^+$	+43.0

2.6.2.   Biological synthesis of metal selenide QDs

Unlike metal sulfide nanoparticles, there are rather few studies on the microbial production of metal selenide nanoparticles (MeSe NPs) (Table 2.4). This is mainly because the microbial production of Se(-II) is challenging and the aqueous selenide (HSe⁻) is rapidly reoxidized to Se(0) under oxic conditions. Therefore, the use of Se oxyanions as precursor material typically requires the addition of a strong reducing agent such as sodium borohydride to produce the required HSe⁻ (Wang et al., 2010; Zheng et al., 2007). There are few microorganisms known that can extend the reduction pathway beyond Se(0) to form HSe⁻ as the end product (Lenz & Lens, 2009; Nancharaiah & Lens, 2015b). An exogenous redox mediator, anthraquinone-2,6-disulfonate (AQDS) has also been used to facilitate the reduction of selenite up to selenide. In the presence of AQDS, a five-fold increase in selenide yield was reported, suggesting the possibility of linking the biosynthesis of QDs precursors to the bioremediation of selenium contaminated waste streams without addition of strong reducing agents (Fellowes et al., 2013).

Table 2.4 Summary of microorganisms, reaction conditions, and characteristics of metal selenide nanoparticles

Microorganism	Precursor materials	Temperature (°C)	pH	NP	Size (nm)	Morphology	Location of synthesis	Ref.
Bacteria								
V. atypica	Cd(ClO$_4$)$_2$, Na$_2$SeO$_3$	37	7.5	ZnSe	30	Spherical	Extra cellular	Pearce et al. (2008)
V. atypica	Cd(ClO$_4$)$_2$, Na$_2$SeO$_3$	37	7.5	CdSe	2.3±1.3	Spherical	Extra cellular	Fellowes et al. (2013)
Pseudomonas sp.	CdCl$_2$, Na$_2$SeO$_3$	37	6.8	CdSe	10-20	-	Intra cellular	Ayano et al. (2014)
E. coli	CdCl$_2$, Na$_2$SeO$_3$	37	-	CdSe	8-11	Spherical	Intra cellular	Suresh, (2014)
Fungi								
H. solani	CdCl$_2$, SeCl$_4$	37	-	CdSe	5.5 ± 2	Spherical	Extra cellular	Suresh, (2014)
S. cerevisae	CdCl$_2$, Na$_2$SeO$_3$	30	-	CdSe	2.69-6.34	Spherical	Intra cellular	Cui et al. (2009)
F. oxysporum	CdCl$_2$, SeCl$_4$	25	-	CdSe	11±2	Spherical	-	Kumar et al. (2007)
Aspergillus niger	-	-	-	PbSe	59 (A.R.: 5-10)	Rod	Extra cellular	Jacob et al. (2014)

2.6.2.1.   MeSe QDs synthesis using bacteria

Formation of MeSe was reported in bacterial cultures for the first time using the selenite reducing bacterium *V. atypica* (Pearce et al., 2008). This microorganism is able to reduce selenite up to HSe⁻, which was used as the precursor for synthesizing CdSe and ZnSe nanoparticles. Addition of heavy metals after the complete reduction of selenite up to selenide was recommended to avoid the formation of a mixture of metal selenide and Se(0) nanoparticles. The formed ZnSe nanoparticles were mainly distributed in the extracellular polymeric substances (EPS) associated with the cells and in the extracellular medium (Pearce et al., 2008) (Fig. 2.4). The drawback of this synthesis route was the larger particle size (30 nm) of the formed ZnSe nanoparticles. These ZnSe particles are not suitable for fluorescence applications as the quantum confinement effect requires ZnSe particles with a diameter below 20 nm. The ZnSe nanoparticles formed were also unstable and became a non-fluorescent precipitate upon exposure to oxygen.

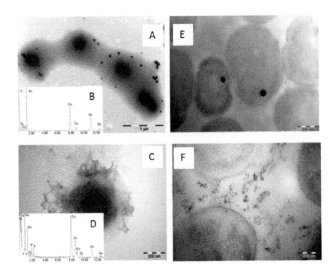

Fig. 2.4 TEM of V. atypica, showing (A) Se⁰ spheres associated with the cells, along with (B) EDX of the particles and also showing (C) ZnSe precipitated outside the cells, along with (D) EDX of the particles. TEM (thin sections) of washed cells of V. atypica, showing (E) intracellular Se⁰ spheres and also showing (F) extracellular ZnSe particles (Adapted from Pearce et al., 2008)

Fellowes et al. (2013) used a similar two-step procedure for the synthesis of CdSe which involved production of biogenic selenide(-II) by *V. atypica* followed. The biogenic Se(-II) was filter sterilized and the pH was increased to 11.2 with NaOH. A solution containing 10 mM $Cd(ClO_4)_2$ and 30 mM reduced glutathione (GSH) was added to the biogenic selenide(-II) for maintaining a final molar ratio of 2:1:3 for Cd:Se:GSH. This approach was successful in forming spherical, smaller and crystalline CdSe particles with diameters below 8 nm (Fellowes et al., 2013).

Biogenic selenide is advantageous as it allows slower growth of metal selenide particles leading to the formation of smaller nanoparticles with monodispersity. Biogenic selenide is more stable and less susceptible to oxidation compared to chemically synthesized selenide (Fellowes et al., 2013). The protein components associated with the biogenic selenide likely contributed to the slower growth rate of the particles and higher stability of the biogenic selenide. One of the proteins from the biogenic Se(-II) solution was identified as the α-subunit of methylmalonyl-CoA decarboxylase, originating from *V. atypica* cells (Fellowes et al., 2013). Another study showed the presence of amide I and amide II bands characteristic of protein molecules associated with the CdSe nanoparticles synthesized by *E. coli* (Yan et al., 2014). The association of proteins on the particle surface can contribute to the biocompatibility of nanoparticles in life science applications.

Recently, Ayano et al. (2014) used a single step procedure for CdSe synthesis. A cadmium resistant selenite reducing *Pseudomonas sp.* strain RB isolated from a soil sample was able to reduce selenite up to selenide in the presence of 1 mM of cadmium. However, separation of large CdSe particles and preventing the excess growth of the particles was necessary as both smaller (~10 nm) and larger (70-100 nm) spherical particles were observed, respectively, inside and outside the cells. The effect of pH, temperature and salinity on CdSe synthesis was investigated. Mesophilic temperature (30°C) and alkaline pH were found to be optimum for CdSe formation. Variation in salinity (from 0.05 to 10 g. L^{-1} NaCl) did not significantly affect CdSe formation. By controlling the selenite and heavy metal concentration, it may be possible to control the growth of the NPs. However, the role of various other factors such as the physiological status and growth phase of bacteria, speciation of heavy metals and incubation time on the MeSe synthesis needs to be investigated in detail (Silver, 1996; Yan et al., 2014).

In our recent paper, a selenium removal mechanism by anaerobic granular sludge in the presence of heavy metals was proposed (Mal et al., 2016). It was revealed that bioreduction of selenite by anaerobic granular sludge was not significantly inhibited in the presence of Pb(II) (150 mg. L^{-1} of Pb) and Zn(II) (400 mg. L^{-1} of Zn). In contrast, selenite reduction was reduced to only 65-48% selenite in the presence of 150 to 400 mg. L^{-1} Cd(II). It also shows that formation of Se(0) or HSe⁻ depends on heavy metals and varied with heavy metal type and its concentration. Hence choosing of heavy metal and optimization of its concentration is necessary for formation of HSe⁻ which precipitates with heavy metals to form metal selenides. However, adsorption of heavy metals onto Se(0) nanoparticles can also be another reason for the partial removal of the heavy metals from aqueous phase . So further speciation studies like X-ray absorption spectroscopic techniques (e.g. X-ray absorption near edge structure (XANES) and Extended X-Ray Absorption Fine Structure (EXAFS)) are required to differentiate, whether the metals are sorbed onto the Se(0) nanoparticles or bound as metal selenides (e.g. CdSe).

2.6.2.2.   **MeSe QDs synthesis using yeast/fungi**

Kumar et al. (2007) demonstrated for the first time production of CdSe nanoparticles using *Fusarium oxysporum*. Fungal mycelium was incubated with selenium tetrachloride (SeCl$_4$) and CdCl$_2$ as selenium and cadmium sources, respectively. Formation of highly stable and monodisperse CdSe quantum dots with a broad fluorescence emission spectrum were observed in the extracellular medium. In a recent study, a plant pathogenic fungus, *Helminthosporum solani* was used for the biosynthesis of CdSe QDs (Suresh, 2014). When the fungal mycelium was incubated with 1 mM of Cd(II) and Se(IV) under ambient conditions, formation of spherical and highly stable CdSe QDs (size 5.5 ± 2.0 nm) in the extracellular medium was observed.

These studies have noted the presence of proteins on CdSe nanoparticles. It was hypothesized that the proteins such as phytochelatins released by the fungus act as capping agents during CdSe nanocrystals growth and help in forming colloids and prevent aggregation (Kumar et al., 2007). The formation of PbSe rods was demonstrated using a marine fungus, *Aspergillus terreus* (Jacob et al., 2014). The extracellularly formed PbSe rods had an average diameter of 57 nm with an aspect ratio between 5 and 10 (Fig. 2.5). However, a more reactive selenium compound, sodium selenosulfate (Na$_2$O$_3$SSe), was used as the precursor material. Similarly, in some of the studies on CdSe synthesis selenium tetrachloride was used as the Se source. The use of selenite or selenate as precursor should be considered different from other studies, because the toxicity and reduction mechanisms of these oxyanions are different from other more reactive selenium compounds like SeCl$_4$ or Na$_2$O$_3$SSe. Since selenium principally exists in the form of oxyanions (selenite or selenate) in wastewaters, the use of oxyanions is preferred if green synthesis of MeSe QDs using inexpensive raw materials and coupled to bioremediation is aimed for.

Baker's yeast, *Saccharomyces cerevisiae*, was used for the production of smaller sized highly fluorescent CdSe QDs (Cui et al., 2009). Monodisperse CdSe nanoparticles with an average diameter of 2 nm were formed inside the cells when CdCl$_2$ was added to selenite reducing yeast cells. CdCl$_2$ was added during the stationary phase to avoid inhibition on growth, glutathione production and selenite reduction. In *S. cerevisiae*, selenite can be reduced by GSH into the selenotrisulfide derivative of glutathione and further reduced into selenocystine. It is suggested that among the selenocompounds, only selenocystine can react with Cd^{2+} and form CdSe QDs. Importantly, the relative activity of GSH-related enzymes in the yeast cells is decreased by addition of 1 mM CdCl₂ slowing down the selenium reduction(Cui et al., 2009). Therefore, the time sequence of adding CdCl₂ is crucial for the successful synthesis of CdSe QDs. Addition of CdCl₂ during the stationary phase of the yeast culture is most suitable to avoid the inhibitory effect of the heavy metal on the reduction of selenite to selenocysteine. Although Cd is generally toxic and high concentrations of Cd inhibit the growth of microorganisms, the addition of 1 mM Cd has been shown to yield smaller sized CdSe nanoparticles (Ayano et al., 2014). Hence, research should focus on finding novel fungal strains with high cadmium resistance for the biosynthesis of CdSe nanoparticles (Ayano et al., 2014).

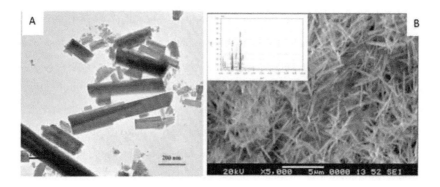

Fig. 2.5 (A) TEM image of the biosynthesized nanorods PbSe. (B) SEM image of the nanorods PbSe; EDAX of the particles also showing biogenic PbSe (inset) (Adapted from Jacob et al., 2014)

2.7. Tellurium based chalcogenides

Te is a P (positive)-type semiconductor and has unique optical and electrical properties. It is an important component in industrial steels, glasses and solar panels (Turner et al., 2012). In the last decade, research on Te has gained considerable interest due to the development of fluorescent CdTe quantum dots with a high quantum yield for *in vivo* cell imaging applications (Deng et al., 2007). A significant amount of research has recently been focused on telluride clusters and nanoparticles as an important tool for new solar cell technology and in biomedicine (Wachter, 2004; Zhang et al., 2007). Te has no known function in living systems. But, microorganisms are involved in the biotransformation of Te oxyanions to insoluble elemental tellurium (Te(0)) or telluride (Te(-II)) (Turner et al., 2012). This bioreduction can be useful in bioremediation efforts of Te polluted wastewaters or soils and couple it to heavy metal removal via metal telluride (MeTe) formation.

2.7.1. Tellurium oxyanion reduction mechanisms

Conversion of Te oxyanion, i.e. tellurate (Te(VI)) or tellurite (Te(IV)) to black elemental tellurium (Te(0)) has been observed in several microorganisms isolated from diverse environments (Turner et al., 2012). Te oxyanions, particularly Te(IV), are extremely toxic to microorganism and it was used as antimicrobial agent in the pre-antibiotic era (Fleming & Young, 1940). Tellurite exerts its toxicity by generating reactive oxygen species in the cytoplasm. So, tellurite uptake is a prerequisite for exerting toxicity (Borsetti et al., 2005; Chasteen et al., 2009; Zannoni et al., 2008). Both the phosphate transporter and acetate permease transporter system may facilitate tellurite uptake by microbial cells. Inside the cells, Te(IV) is rapidly detoxified by reduction to elemental Te(0) precipitates, either by membrane-bound nitrate reductase (Sabaty et al., 2001) or by the thioredoxin (Trx)-glutathione (GSH) system (Zannoni et al., 2008). Tellurite reduction observed in various microorganisms is mediated through respiratory or detoxification mechanisms[114]. Dissimilatory reduction of the Te(IV)/Te(0) redox couple is thermodynamically favourable for anaerobic respiration. But, evidence on the coupling of Te(IV) reduction and microbial growth is not substantial (Baesman et al., 2007).

2.7.2. Biological synthesis of cadmium telluride QDs

Te(IV) reduction to Te(0) was reported in different bacterial cultures such as *R. capsulatus* (Borsetti et al., 2003), *R. sphaeroides* (Moore & Kaplan, 1994), *Pseudomonas pseudoalcaligenes* KF707 (di Tomaso et al., 2002) and *B. selenitireducens* (Baesman et al., 2007). But information on reduction of tellurite beyond elemental Te and the formation of telluride is limited. Nevertheless, formation of stable CdTe nanoparticles was observed in bacterial and fungal cultures (Table 2.5).

2.7.2.1. CdTe QDs synthesis using bacteria

Bao et al. (2010) were the first to report biogenic synthesis of CdTe quantum dots using *E. coli* K12 cells. In the synthetic procedure, microbial cells were incubated with tellurite and

CdCl$_2$ along with a strong reducing agent, sodium borohydride. Formation of CdTe nanoparticles was observed in the extracellular medium and the reduction employed in this procedure is analogous to the chemical synthesis, microbial cells acted only as sites for complexation of metal ions and growth of CdTe nuclei. Interestingly, CdTe nanocrystals with tunable size-dependent emission from blue to green were produced by controlling the incubation time (Bao et al., 2010). With prolongation of the incubation time, the absorption edge and the photoluminescence maxima of the biosynthesized CdTe QDs shifted towards longer wavelengths, due to the increase in particle size.

Table 2.5 Summary of microorganisms, reaction conditions and characteristics of biogenic CdTe nanoparticles

Microorganism	Precursor materials	Temperature (°C)	pH	Size (nm)	Morphology	Location of synthesis	Ref.
Bacteria							
Bacillus pumilus	CdI$_2$, Na$_2$TeO$_3$	30	7.0	6-10	Spherical	Extra cellular	Pawar et al. (2013)
E. coli	CdCl$_2$, K$_2$TeO$_3$	-	-	6	Spherical	Intra cellular	Monras et al. (2012)
E. coli	CdCl$_2$, Na$_2$TeO$_3$	37	7.0	2.0-3.2	Spherical	Extra cellular	Bao et al. (2010)
Fungi							
F. oxysporum	CdCl$_2$, TeCl$_4$	25	7.0	15-20	Spherical	Extra cellular	Syed & Ahmad, (2013)
Serratia marcescens	CdCl$_2$, Na$_2$TeO$_3$	30	7.0	2-3.6	Spherical	Extra cellular	Pawar et al. (2013)
S. cerevisiae	CdCl$_2$, Na$_2$TeO$_3$	35	-	2-3.6	Spherical	Extra cellular	Bao et al. (2010)

The Ostwald ripening process was used to explain the growth and formation of crystalline CdTe quantum dots in microbial systems (Bao et al., 2010). Upon exposure to metal ions, *E. coli* cells produce metal binding proteins as part of a stress response mechanism and secrete them into the extracellular medium. The proteins secreted by bacterial cells bind to the CdTe nuclei and crystals and stabilize them as colloidal nanocrystals (Fig. 2.6). Association of proteins with CdTe improves the biocompatibility of the CdTe nanoparticles (Bao et al., 2010). However, the mechanism of formation and growth of CdTe nanocrystals, particularly the nature and origin of proteins involved in the nanocomposite formation in microorganisms are poorly understood.

GSH, an abundant thiol in microorganisms, has a dual role in nanoparticle synthesis. It can act as both reducing and stabilizing agent. GSH mediates the reduction of selenite and tellurite to Se(0) and Te(0), respectively. The role of GSH in the formation of Te(-II) from tellurite was studied in *E. coli* cells (Perez-Donoso et al., 2012). *E. coli* strains overexpressing gshA and gshB genes involved in glutathione synthesis was used for the production CdTe QDs (Monra's et al., 2012). Despite the fact that gshA and gshB genes are both involved in GSH synthesis, GSH production was much higher in case of the gshA overexpressing *E. coli* strain which eventually produced CdTe QDs, but not in case of the wild or gshB overexpressing *E. coli* strain. Similar to the chemical synthesis of CdTe QDs, it is clear that a bacterial strain containing increased levels of GSH should be a good candidate for biosynthesizing this kind of NPs (Monra's et al., 2012).

Increasing the incubation temperature (42°C) leads to the enhancement of the fluorescence of the CdTe QDs, while incubation of bacteria with citrate buffer changes the fluorescence color, suggesting that NPs with different spectroscopic properties can be produced by varying the incubation temperature as well as the buffer (Monra's et al., 2012). This could be the consequence of size, shape and/or composition changes mediated by a specific cellular status or factor. However, the authors did not report the exact reason behind the change in fluorescence properties of CdTe QDs. Interestingly, significant effects of increased temperature offers a new horizon of using thermophiles to improve biological synthesis of QDs.

The optimal conditions and factors to enhance NP biosynthesis by microorganisms are far from being understood. For example, although no proper explanation was found, it was reported that CdI_2 was required for CdTe synthesis by a *B. pumilus*, while *Serratia marcescens* needed $CdCl_2$. This suggests proper cadmium salts are required for the synthesis of CdTe depending on the microorganism used (Pawar et al., 2013). However, more research on the cellular events underlying the biomolecular mechanism(s) involved in the bacterial production of CdTe QDs will allow the bacterial production of CdTe QDs with defined properties.

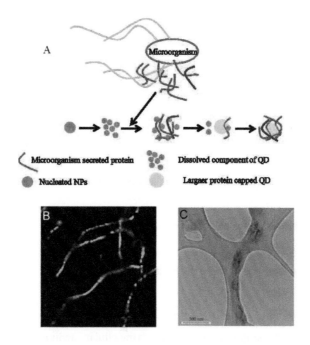

Fig. 2.6 A) Schematic diagram of the mechanism for CdTe QDs biosynthesis. B) Confocal image of E. coli after 7 days incubation. C) TEM image of the synthesized CdTe QDs on bacterial surface after 7 days incubation (Adapted from Bao et al., 2010)

2.7.2.2.   CdTe QDs synthesis using fungi/yeast

Bao et al. (2010) demonstrated a simple and efficient biosynthesis of CdTe QDs by using *Saccharomyces cerevisiae*. Highly fluorescent CdTe QDs with a particle size ranging from 2.0 to 3.6 nm were well dispersed in the cytoplasm and nucleus of the yeast cells. Although NaBH$_4$ was used as a strong reducing agent in this protocol, yeast cells played a key role in secreting proteins which control the size and fluorescence properties of CdTe QDs. Proteins secreted by the yeast cells as part of a defense strategy in response to Cd and Te play a role in the formation and stabilization of CdTe colloids. Two proteins were identified with 7.7 and 692 kDa from the CdTe nanoparticles (Bao et al., 2010). Interestingly, the quantum yield of the biogenic CdTe QDs was ~33%, which is higher than what is achieved for high quality CdTe nanocrystals using a hydrothermal procedure performed at a temperature of 180°C (Bao et al., 2010). This shows that the formation of high quality CdTe QDs suitable of fluorescence applications using biological systems is well possible.

Syed and Ahmad (2013) demonstrated the synthesis of highly fluorescent extracellular CdTe QDs using the fungal strain *F. oxysporum* without using any chemical reducing agent. $CdCl_2$ and $TeCl_4$ were used as Cd and Te precursors, respectively, incubated with the fungal mycelium at room temperature for 96 h on a rotary shaker (200 rpm). Highly stable and water dispersible CdTe nanoparticles with a size ranging from 15 to 20 nm were produced. Given the growing knowledge on the interaction mechanisms between fungi and metal ions (Te/Cd), along with the important and desired applications in medicine, the field of microbially driven synthesis of CdTe QDs is expected to become a very active research area.

2.8. Mechanisms of biological synthesis of metal chalcogenides

2.8.1. Localization of metal chalcogenides

High resolution microscopy of MeCh synthesizing microorganisms showed the presence of MeCh in their cytoplasm, periplasm and in their extracellular medium. Although, the exact mechanisms and sites of MeCh synthesis in microorganisms are still unknown, based on microscopic evidence, the formation of MeCh in the cytoplasm and outside the cells is hypothesized (Fig. 2.7). Since the particle size is small, their transport across the cell membrane from the cytoplasm to the extracellular medium and vice versa was considered. Sweeney et al. (2004) demonstrated that when *E. coli* cells were incubated with CdS particles, the majority of the added particles were seen outside the cells suggesting a minimal transport of nanoparticles across the cell membrane. This observation only shows that the transport of externally added particles to the cytoplasm is limited, but does not exclude transport mechanisms of intracellularly formed particles, if any.

A study on the microbial synthesis of CdS by *R. palustris* reported a totally opposite phenomenon, wherein CdS nanoparticles were biosynthesized in the cytoplasm and transported to the extracellular medium (Bai & Zhang, 2009). ZnSe nanoprecipitates formed by *V. atypica* were observed in the medium and in the EPS surrounding the cells suggesting the formation of nanomaterials directly in the extracellular medium (Pearce et al., 2008).

2.8.2.   Size and shape of biogenic metal chalcogenides

Control of the particle size and distribution during the synthesis of nanoparticles is one of the most important criteria as applications of metal chalcogenide nanoparticles are mostly influenced by their size and shape. By controlling the environmental parameters at which nanoparticles are synthesized, control over the sizes and shapes during the synthesis of nanoparticles can be achieved (Gurunathan et al., 2009). Pandian et al. (2011) demonstrated that at pH 9, *Brevibacterium casei* SRKP2 showed maximum synthesis of smaller sized CdS NPs, when compared to other pH investigated. The pH of the solution highly influences the reduction reaction of metal(loid)s. Since, the precipitation rate is inversely proportional to the proton concentrations; a higher pH increases the precipitation rate, which eventually facilitates nuclei formation and growth of smaller sized nanocrystals. The effect of the medium pH on the size of the nanocrystals could also be due to a balance between nucleation and crystal growth.

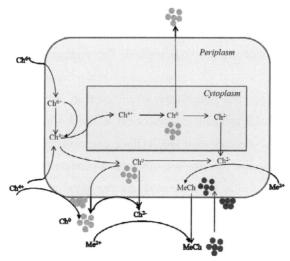

Fig. 2.7 Mechanisms of biological synthesis of selenium and metal selenide nanoparticles

It is suggested that MeCh QDs with different size can be achieved by controlling the incubation time (Cui et al., 2009). With the increase in incubation time of yeast cells with $CdCl_2$ from 10 to 40 h, the average size of CdSe NPs increased from 2.6 to 6.4 nm (Cui et al., 2009), while Moon et al. (2014) reported an increase in mean particle diameter with a broader particle size distribution when the incubation time of the microbial synthesis of ZnS was increased from 1 to 5 d.

Moon et al. (2014) showed that addition of a pulse dose of 1 mM Zn(II) day^{-1} to a culture medium containing 10 mM thiosulfate produced 12 nm sized ZnS particles, while smaller ZnS particles (6.5 nm) were formed with a pulse dose of 1 mM Zn(II) day^{-1} to 5 mM thiosulfate. An increase in precursor concentration from 0.5 to 10 mM (Cd^{2+} or S^{2-}) caused an increase in the average particle size and a consequent shift towards red color in the emission wavelength (Mi et al., 2011). So, like the chemical synthesis, the concentration and molar ratio of precursor materials can affect the particle size of the MeCh QDs formed in microbial cultures.

2.9. Applications of chalcogenide QDs

The application of MeCh QDs, as a new technology for biosystems, has been mostly studied in mammalian cells. With the introduction of water-soluble and bioconjugated CdSe–ZnS QDs to cell labeling and imaging by Chan and Nie (1998) and (Alivisatos, 1996), the applications of QDs to approach biological problems received new momentum. There are now various advanced detection methods available for tracking and detection of QDs in living systems including confocal microscopy, total internal reflection microscopy, wide-field epifluorescence microscopy and fluorometry, which show an increasing tendency to apply QDs as markers in medicine and biology (Bruchez et al., 1998; Chan & Nie, 1998; Michalet et al., 2005). Some of the applications of MeCh QDs are summarized in Fig. 2.8. However, as impressive as this is, toxicity concerns are still valid for longer periods and for biological applications (Chen et al., 2012). In the demand of using biocompatible and less toxic QDs, new developments include polyethylene glycol (PEG) coating or peptide conjugation for cytosol localization and nucleus targeting (Chen & Gerion, 2004; Yildiz et al., 2009), and increased focus on the biological synthesis of less toxic and biocompatible MeCh QDs.

2.9.1. Cell imaging and tracking

2.9.1.1. *In-vitro* Imaging

QDs are used as superior fluorescent labels for visualizing cells in *in vitro* assays. For instance, water soluble PbS and PbSe were used to label human colon cancer cells for imaging cancerous cells (Hyun et al., 2007). Similarly, dihydrolipoic acid-capped CdSe and

ZnS QDs were used for labelling HeLa cells (Jaiswal et al., 2003). Li et al. (2007) demonstrated the labeling of *Salmonella typhimurium* cells by using 3-mercaptopropionic acid (3-MPA) capped CdS. This approach allows development of an economic imaging method using CdS QDs for detection of *S. typhimurium* cells for practical applications[136]. Moreover, with the help of surface modifications of MeCh QDs, the application of MeCh QDs can be increased vastly *in vitro* and *in vivo* imaging. Chen and Gerion (2004) demonstrated the application of CdSe/ZnS nanocrystal-peptide conjugates as non-toxic, long-term imaging tool for observing nuclear trafficking mechanisms and cell nuclear processes.

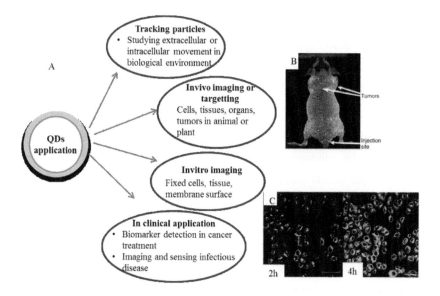

Fig. 2.8 A) Applications of QDs in medicine and biology, B) Fluorescence image of human prostate cancer implanted in a mouse. The tumor is targeted with anti-PSMA antigen conjugated CdSe/ZnS QDs (Adopted from Medintz et al., 2005). C) Cellular internalization efficiency of silica-coated CdSe QDs in cancer using CLSM study; QDs were incubated in presence of HeLa cells. The CLSM micrographs confirmed the peak internalization of QDs at 4 h (Adapted from Vibin et al., 2014)

The visualization of the movement of the CdSe/ZnS nanocrystal–peptide conjugates from the cytoplasm to the nucleus as well as the accumulation of the complex in the cell nucleus over a long observation time period is also possible (Chen & Gerion, 2004). Water-soluble, biologically compatible CdSe QDs with L-cysteine as capping agent has been used to label serum albumin (BSA) and *E. coli* cells (Liu et al., 2009). Long-term live cell imaging to track

39

whole cells or intracellular biomolecules by QDs is also possible and was demonstrated by Hasegawa et al. (2005). Jaiswal et al. (2003) also demonstrated that labeling living cells using CdSe/ZnS QDs does not affect the growth of normal cells and cell signaling. Importantly, the fluorescence performance of MeCh QDs in the cells can last over a week and does not affect cell morphology.

2.9.1.2. In-vivo imaging

MeCh QDs have attracted interest for their applications for *in-vivo* imaging (Choi et al., 2010; Hamada et al., 2011; Weng et al., 2008). In vivo application requires MeCh QDs with low toxicity, high contrast, high sensitivity and photostability. In 2002, Akerman et al. (2002) reported the *in vivo* application of MeCh QD by injecting a peptide coated CdSe–ZnS QD into the tail vein of a mouse and demonstrated the specificity of the conjugate to endothelial cells in the lung blood vessels. During the same time period, Dubertret et al. (2002) demonstrated *in vivo* application of micelle-coated CdSe/ZnS QDs for fluorescent imaging of *Xenopus* embryos by microinjecting QD into a *Xenopus* embryo. These two *in vivo* experiments brought radical changes in the biological applications of MeCh QDs. Tumor targeting for early diagnosis of cancer by using CdSe/CdS/ZnS quantum rods (QRs) coated with PEGylated phospholipids and arginine-glycine-aspartic acid (RGD) peptide is possible and now offering new opportunities for imaging of early tumor growth. Till now, many reports are available on *in vivo* applications such as locating draining lymph nodes, visualizing blood cells (Alivisatos et al., 2005; Crookes-Goodson et al., 2008), targeting vasculatures, and imaging tumors (Ballou et al., 2007; Cai & Chen, 2008; Chapman et al., 2005; Gao et al., 2010b; Kairdolf et al., 2013; Li et al., 2014; Walling et al., 2009).

2.9.1.3. Cell tracking

QDs are insensitive to photobleaching even after prolonged exposure to light sources as opposed to organic fluorophores which makes it easier to track a particular stained cell. For specific cell labeling, imaging and tracking by using QDs, functionalization of QDs with biomolecules is a critical step. Functionalized QDs have been frequently used for multi-color imaging, through which multiple targets could be simultaneously detected and profiled (Delehanty et al., 2011; Kobayashi et al., 2007). It has also been extensively applied to cell surface receptor labeling (Ag et al., 2014; Jeyadevi et al., 2013; Zrazhevskiy & Gao, 2013),

40

intracellular biomolecule tracking and sensing (Biermann et al., 2014; Zhang et al., 2013) and organelle targeting (Xu et al., 2012). Recent progress of QD-based cell imaging and tracking includes multiplex protein tracking, monitoring intracellular protein interaction dynamics, stem cell labeling and imaging as well as detection of gene expression (Biermann et al., 2014; Chen et al., 2012a; McConnachie et al., 2013; Rasmussen et al., 2013). Moreover, it is also quite useful for single molecule imaging and tracking, which allows to follow a single molecule in real time and to visualize the actual molecular dynamics in their habitat environment (Chang & Rosenthal, 2011; Clausen et al., 2014).

2.9.2. Cancer imaging

The use of MeCh QDs for cancer imaging is one of the most promising applications of QDs (Fang et al., 2012; Vibin et al., 2014). One classical example is the localization of a QD-secondary antibody conjugated to Her2 (hairy-related 2 protein) (Wu et al., 2003). Conjugates of antibodies and peptides with QDs have received much attention as potential markers of various cancers. Gao et al. (2004) successfully demonstrated imaging of prostate cancer in nude mice by using a MeCh QD conjugated with an antibody raised for prostate specific membrane antigen (PSMA). Akerman et al. (2002) showed application of ZnS-capped CdSe QDs conjugated with specific peptides for *in vitro* and *in vivo* targeted imaging of lung endothelial cells, brain endothelial cells, and breast carcinoma cells. QD-assisted mapping of lymph nodes has been considered a promising technique for staging certain types of cancers (Akerman et al., 2002).

Kim et al. (2004) reported fluorescence imaging of lymph nodes in animal models using CdTe/CdSe core/shell type QDs (Kim et al., 2004). The near-infrared (NIR) imaging employing MeCh QDs with accessibility to distant lymph nodes and specificity to lymph node metastasis is promising for image-guided pre-surgical and surgical oncology of gastrointestinal tumors, metastasis of spontaneous melanoma, breast cancer and non-small cell lung cancer (Choi et al., 2010; Wang et al., 2015; Xue et al., 2012a; Yuan et al., 2014). MeCh QDs with high quantum yields can be used as labels to improve imaging of fluorescence in situ hybridization (FISH) analysis of human chromosomal changes. Xiao and Barker have investigated coated (CdSe)ZnS QDs as fluorescence labels for FISH of biotinylated DNA to human lymphocyte metaphase chromosomes and demonstrated the detection of the clinically relevant HER2 locus in breast cancer cells by FISH (Xiao &

Barker, 2004). Recently, Yang et al. (2013) demonstrated that it is possible to link anti-epidermal growth factor receptor (EGFR) antibodies with CdSeTeS QDs modified with alpha-thioomega-carboxy poly(ethylene glycol). Moreover, these newly synthesized quaternary-alloyed QDs showed significantly long fluorescence lifetimes (> 100 ns) as well as excellent photostability. The authors also showed that modified CdSeTeS QDs with the EGFR antibodies can be applied as labeling probes for targeted imaging of EGFR on the surface of SiHa cervical cancer cells through conjugation of QDs with the anti-EGFR antibodies.

2.9.3.   Cytotoxicity of MeCh QDs

With the rapid development in the synthesis and commercialization of MeCH QDs, their release into the environment is also inevitable. This may pose hazards to ecosystem well-being and human health (Kirchner et al., 2005; Lovrić et al., 2005; Wiecinski et al., 2013). One of the main reasons for QD cytotoxicity is the desorption of Cd (i.e. QD core degradation), free radical formation, interaction with intracellular components or bioavailability (uptake) of QDs (Jamieson et al., 2007). Exposure of the CdSe core to an oxidative environment can cause decomposition and desorption of Cd ions, which plays an important role in subsequent toxicity. The generation of intracellular reactive oxygen species (ROS) has also been shown to be a controlling factor of the toxicity (Valizadeh et al., 2012). Although Cd can generate free radicals, it is not clear whether or not the generation of free radicals depends on Cd desorption from QDs (Oh & Lim, 2006). In addition to the effects of the MeCh QD core components, ligands or the surface coating added to the core MeCH QDs to stabilize and make it biologically active may also exert toxic effects on cells. Mercaptopropionic acid and mercaptoacetic acid are mildly cytotoxic (Kirchner et al., 2005), while mercaptoundecanoic acid and TOPO have the ability to damage DNA in the absence of the QD core (Hoshino et al., 2004).

In one of our studies, biogenic nano-Se (nano-Se[b]) synthesized by anaerobic granular sludge was 10-fold less toxic than chemically synthesized nano-Se (nano-Se[c]) (**Chapter 8**). The differences in toxicity can be due to the presence of different surface stabilizing agents of nano-Se[b] and nano-Se[c]. Nano-Se[c] is stabilized by a single protein, BSA, while nano-Se[b] is stabilized by extracellular polymeric substances (EPS) present on the surface of nano-Se[b], originating from the microorganisms present in the anaerobic granular sludge. It is suggested that the presence of EPS increases the physiochemical stability of NPs and prevents their dissolution (Liu & Hurt, 2010; Moreau et al., 2007).

The presence of humic acids can inhibit the generation of intracellular ROS, which could be responsible for the lower toxicity of NPs (Lin et al., 2012). Recently, Bondarenko et al. (2016) showed that levan (a fructose-composed biopolymer of bacterial origin) as a surface coating significantly reduced the toxic effects of Se-NPs in an in vitro assay on the human cell line Caco-2 (colorectal adenocarcinomatous tissue of the human colon). There are clear indications that the presence of EPS of bacterial origin can alleviate the toxicity of the NPs and more emphasis should be given to the use of biological synthesis for commercial production of MeCh QDs. Also, there is no need to use a hazardous surface stabilizer e.g. TOPO when it is formed by a biological route.

Groups III–V QDs have a lower cytotoxicity and may provide a more stable alternative to groups II–VI QDs (Bharali et al., 2005). However, these QDs tend to have much lower quantum efficiencies and synthesis of these QDs is also difficult on a competitive time scale. Moreover, as described above, toxicity of QDs not only depends on its core (i.e. Cd), but on several other factors. In contrast, the use of microorganisms for production of MeCh QDs not only provides a low cost, environmentally friendly method, it also combines the bioremediation approaches, i.e. chalcogen oxyanion reduction, which tackle natural processes to convert environmentally toxic wastes into less toxic forms. With the help of the biological MeCh QDs synthesis route, it is possible to convert industrial waste streams to 'high-end' industrial products like MeCh QDs.

References

Ag, D., Bongartz, R., Dogan, L.E., Seleci, M., Walter, J.G., Demirkol, D.O., Stahl, F., Ozcelik, S., Timur, T. 2014. Biofunctional quantum dots as fluorescence probe for cell-specific targeting. *Colloids Surf B.*, **1**(114), 96-103.

Aguiera-Sigalat, J., Rocton, S., Sánchez-Royo, J.F., Galian, R.E., Pérez-Prieto, J. 2012. Highly fluorescent and photostable organic- and water-soluble CdSe/ZnS core shell quantum dots capped with thiols. *RSC Adv.*, **2**, 1632-1638.

Akerman, M.E., Chan, W.C., Laakkonen, P., Bhatia, S.N., Ruoslahti, E. 2002. Nanocrystal targeting in vivo. *Proc Natl Acad Sci USA*, **99**, 12617-12621.

Al-Shalabi, Z., Stevens-Kalceff, A.M., Doran, M.P. 2014. Application of *Solanum lycopersicum* (tomato) hairy roots for production of passivated CdS nanocrystals with quantum dot properties. *Biochem Eng J.*, **84**, 36-44.

Alivisatos, A.P. 1996. Semiconductor clusters, nanocrystals, and quantum dots. *Science*, **271**, 933-937.

Alivisatos, A.P., Gu, W., Larabell, C. 2005. Quantum dots as cellular probes. *Annu Rev Biomed Eng.*, **7**, 55-76.

Ayano, H., Kuroda, M., Soda, S., Ike, M. 2014. Effects of culture conditions of *Pseudomonas aeruginosa* strain RB on the synthesis of CdSe nanoparticles. *J Biosci Bioeng.*, **119**(4), 440-445.

Ayano, H., Miyake, M., Terasawa, K., Kuroda, M., Soda, S., Sakaguchi, T., Ike, M. 2013. Isolation of a selenite-reducing and cadmium-resistant bacterium *Pseudomonas sp.* strain RB for microbial synthesis of CdSe nanoparticles. . *J Biosci Bioeng.*, **117**(5), 576-581.

Azabou, S., Mechichi, T., Patel, B., Sayadi, S. 2007. Isolation and characterization of a mesophilic heavy-metals-tolerant sulfate-reducing bacterium *Desulfomicrobium* sp. from an enrichment culture using phosphogypsum as a sulfate source. *J Hazard Mater.*, **140**((1-2)), 264-270.

Azzazy, M.E.H., Mansour, M.H.M., Kazmierczak, C.S. 2007. From diagnostics to therapy: Prospects of quantum dots. *Clinical Biochemistry*, **40**, 917-927.

Baesman, S.M., Bullen, T.D., Dewald, J., Zhang, D.H., Curran, S., F.S., I. 2007. Formation of tellurium nanocrystals during anaerobic growth of bacteria that use Te oxyanions as respiratory electron acceptors. . *Appl Environ Microbiol.*, **73**, 2135-2143.

Bai, H.J., Zhang, Z.M. 2009. Microbial synthesis of semiconductor lead sulfide nanoparticles using immobilized *Rhodobacter sphaeroides*. *Mater Lett.*, **63**, 764-766.

Bai, H.J., Zhang, Z.M., Gong, J. 2006. Biological synthesis of semiconductor zinc sulfide nanoparticles by immobilized *Rhodobacter sphaeroides*. *Biotechnol Lett.*, **28**, 1135-1139.

Bai, J.H., Zhang, M.Z., Guo, Y., Yang, E.G. 2009. Biosynthesis of cadmium sulfide nanoparticles by photosynthetic bacteria *Rhodopseudomonas palustris*. *Colloids Surf B.*, **70**, 142-146.

Ballou, B., Ernst, L.A., Andreko, S., Harper, T., Fitzpatrick, J.A.J., Waggoner, A.S., Bruchez, M.P. 2007. Sentinel lymph node imaging using quantum dots in mouse tumor models. *Bioconjugate Chem.*, **18**, 389-396.

Bao, H., Hao, N., Yang, Y., Zhao, D. 2010. Biosynthesis of Biocompatible Cadmium Telluride Quantum Dots Using Yeast Cells. *Nano Res.*, **3**, 481-489.

Baumle, M., Stamou, D., Segura, J.M., Hovius, R., Vogel, H. 2004. Highly fluorescent streptavidin-coated CdSe nanoparticles: preparation in water, characterization, and micropatterning. *Langmuir*, **20**, 3828-3831.

Benedetto, S.J., de Almeida, K.S., Gomes, A.H., Vazoller, F.R., Ladeira, A.C.Q. 2005. Monitoring of sulfate-reducing bacteria in acid water from uranium mines. *Min Eng.*, **18**, 1341-1343.

Bharali, D.J., Lucey, D.W., Jayakumar, H., Pudavar, H.E., Prasad, P.N. 2005. Folate-receptor-mediated delivery of InP quantum dots for bioimaging using confocal and two-photon microscopy. *J Am Chem Soc.*, **127**, 11364-11371.

Biermann, B., Sokoll, S., Klueva, J., Missler, M., Wiegert, J.S., Sibarita, J.B., Heine, M. 2014. Imaging of molecular surface dynamics in brain slices using single-particle tracking. *Nat Commun.*, **5**, 3024.

Biju, V., Itoh, T., Anas, A., Sujith, A., Ishikawa, M. 2008. Semiconductor quantum dots and metal nanoparticles: syntheses, optical properties, and biological applications. *Anal Bioanal Chem.*, **391**(7), 2469-2495.

Bondarenko, O.M., Ivask, A., Kahru, A., Vija, H., Titma, T., Visnapuu, M., Joost, U., Pudova, K., Visnapuu, T., Alamäe, T. 2016. Bacterial polysaccharide levan as stabilizing, non-toxic and functional coating material for microelement-nanoparticles. *Carbohydr Polym.*, **20**, 710-720.

Borsetti, F., Borghese, R., Francia, F., Randi, M.R., Fedi, S., Zannoni, D. 2003. Reduction of potassium tellurite to elemental tellurium and its effect on the plasma membrane redox components of the facultative phototroph *Rhodobacter capsulatus*. *Protoplasma*, **221**, 152-161.

Borsetti, F., Tremaroli, V., Michelacci, F., Borghese, R., Winterstein, C., Daldal, F., Zannoni, D. 2005. Tellurite effects on *Rhodobacter capsulatus* cell viability and superoxide dismutase activity under oxidative stress conditions. . *Res Microbiol.*, **156**, 807-813.

Bruchez, M., Moronne, M., Gin, P., Weiss, S., Alivisatos, A.P. 1998. Semiconductor nanocrystals as fluorescent biological labels. *Science*, **281**, 2013-2016.

Cai, W.B., Chen, X.Y. 2008. Preparation of peptide-conjugated quantum dots for tumor vasculature-targeted imaging. *Nat Protoc.*, **3**, 89-96.

Castuma, C.E., Huang, R., Kornberg, A., Reusch, R.N. 1995. Inorganic polyphosphates in the acquisition of competence in *Escherichia coli. J Biol Chem.*, **270**, 12980-12983.

Chan, W.C.W., Nie, S.M. 1998. Quantum dot bioconjugates for ultrasensitive nonisotopic detection. *Science.*, **281**, 2016-2018.

Chang, J.C., Rosenthal, S.J. 2011. Real-time quantum dot tracking of single proteins. *Methods Mol Biol.*, **726**, 51-62.

Chang, Y.J., Peacock, A.D., Long, P.E., Stephen, J.R., McKinley, J.P., Macnaughton, S.J., Hussain, A.K., Saxton, A.M., White, D.C. 2001. Diversity and characterization of sulfate-reducing bacteria in groundwater at a uranium mill tailings site. *Appl Environ Microbiol.*, **67**, 3149-3160.

Chang, Y.P., Pinaud, F., Antelman, J., Weiss, S. 2008. Tracking bio-molecules in live cells using quantum dots. *J Biophotonics.*, **1**(4), 287-298.

Chapman, S., Oparka, K.J., Roberts, A.G. 2005. New tools for in vivo fluorescence tagging. *Curr Opin Plant Biol.*, **8**, 565-573.

Chasteen, T.G., Fuentes, D.E., Tantaleán, J.C., Vásquez, C.C. 2009. Tellurite: history, oxidative stress, and molecular mechanisms of resistance. . *FEMS Microbiol Rev.*, **33**, 820-832.

Chen, F.Q., Gerion, D. 2004. Fluorescent CdSe/ZnS nanocrystal-peptide conjugates for long-term, nontoxic imaging and nuclear targeting in living cells. *Nano Lett.*, **4**(10), 1827-1832.

Chen, G., Yi, B., Zeng, G., Niua, Q., Yan, M., Chen, A., Dua, J., Huanga, J., Zhang, Q. 2014. Facile green extracellular biosynthesis of CdS quantum dots by white rot fungus *Phanerochaete chrysosporium. Colloids Surf B.*, **117**, 199-205.

Chen, L.Q., Xiao, S.J., Hu, P.P., Peng, L., Ma, J., Luo, L.F., Li, Y.F., Huang, C.Z. 2012a. Aptamer-mediated nanoparticle-based protein labeling platform for intracellular imaging and tracking endocytosis dynamics. *Anal Chem.*, **84**(7), 3099-3110.

Chen, N., He, Y., Su, Y., Li, X., Huang, Q., Wang, H., Zhang, X., Tai, R., Fan, C. 2012. The cytotoxicity of cadmium-based quantum dots. *Biomaterials*, **33**, 1238-1244.

Choi, S.H., Liu, H.W., Liuetal, B.F. 2010. Design considerations for tumour-targeted nanoparticles. *Nat Nanotechnol.*, **5**(1), 42-47.

Clausen, M.P., Arnspang, E.C., Ballou, B., Bear, J.E., Lagerholm, B.C. 2014. Simultaneous Multi-Species Tracking in Live Cells with Quantum Dot Conjugates. *PLoS ONE*, **9**(6), 97671.

Crookes-Goodson, W.J., Slocik, J.M., Naik, R.R. 2008. Bio-directed synthesis and assembly of nanomaterials. *Chem Soc Rev.*, **37**(11), 2403-2412.

Cui, R., Liu, H.H., Xie, H.Y., Zhang, Z.L., Yang, Y.R., Pang, D.W., Xie, Z.X., Shen, P. 2009. Living yeast cells as a controllable biosynthesizer for fluorescent quantum dots. *Adv Funct Mater.*, **19**(15), 2359-2364.

da Costa, P.J., Girão, V.A., Lourenço, P.J., Monteiro, O.C., Trindade, T., Costa, M.C. 2012. Synthesis of nanocrystalline ZnS using biologically generated sulfide. *Hydrometallurgy*, **117-118** 57-63.

Dameron, C.T., Reese, R.N., Mehra, R.K., Kortan, A.R., Brus, L.E., Winge, D.R. 1989. Biosynthesis of cadmium sulphide quantum semiconductor crystallites. *Nature*, **338**, 596-597.

Delehanty, J.B., Bradburne, C.E., Susumu, K., Boeneman, K., Mei, B.C., Farrell, D., Blanco-Canosa, J.B., Dawson, P.E. 2011. Spatiotemporal Multicolor Labeling of Individual Cells Using Peptide-Functionalized Quantum Dots and Mixed Delivery Techniques. *J Am Chem Soc.*, **133**(27), 10482-10489.

Deng, Z.T., Zhang, Y., Yue, J.C., Tang, F.Q., Wei, Q. 2007. Green and orange CdTe quantum dots as effective pH-sensitivefluorescent probes for dual simultaneous and independent detection of viruses. *J Phys Chem. B*, **111**, 12024-12031.

di Tomaso, G., Fedi, S., Carnevali, M., Manegatti, M., Taddei, C., Zannoni, D. 2002. The membrane--bound respiratory chain of *Pseudomonas pseudoalcaligenes* KF707 cells grown in the presence or absence of potassium tellurite. . *Microbiology*, **148**, 1699-1708.

Dubertret, B., Skourides, P., Norris, D.J., Noireaux, V., Brivanlou, A.H., Libchaber, A. 2002. In vivo imaging of quantum dots encapsulated in phospholipid micelles. *Science*, **298**(5599), 1759-1762.

Edwards, C., Beatty, C.J., Loiselle, J., Vlassov, K., Lefebvre, D.D. 2013. Aerobic transformation of cadmium through metal sulfide biosynthesis in photosynthetic microorganisms. *BMC Microbiology*, **13**(161), 1-11.

El-Shanshoury, A., Elsilk, E.S., Ebeid, E.M. 2012. Rapid biosynthesis of cadmium sulfide (CdS) nanoparticles using culture supernatants of *Escherichia coli* ATCC 8739, *Bacillus subtilis* ATCC 6633 and *Lactobacillus acidophilus* DSMZ 20079T. *Afr J Biotechnol.*, **11**(31), 7957-7965.

Fang, M., Peng, C.W., Pang, D.W., Li, Y. 2012. Quantum Dots for Cancer Research: Current Status, Remaining Issues, and Future Perspectives. *Cancer Biol Med.*, **9**(3), 151-163.

Fellowes, J.W., Pattrick, R.A.D., Lloyd, J.R., Charnock, J.M., Coker, V.S., Mosselmans, W., Weng, T.C., Pearce, C.I. 2013. Ex situ formation of metal selenide quantum dots using bacterially derived selenide precursors. *Nanotechnology*, **24**(14), 145603-145612.

Fleming, A., Young, M.Y. 1940. The inhibitory action of potassium tellurite on coliform bacteria. *J Pathol Bacteriol.*, **51**, 29-35.

Gallardo, C., Monrása, P.J., Plaza, O.D., Collao, B., Saona, L.A., Durán-Toroa, V., Venegas, F.A., Pérez-Donoso, J.M. 2014. Low-temperature biosynthesis of fluorescent semiconductor nanoparticles (CdS) by oxidative stress resistant Antarctic bacteria. *J Biotechnol.*, **187**, 108-115.

Gallegos-Garcia, M., Celis, L.B., Rangel-Méndez, R., Razo-Flores, E. 2009. Precipitation and recovery of metal sulfides from metal containing acidic wastewater in a sulfidogenic down-flow fluidized bed reactor. *Biotechnol Bioeng.*, **102**(1), 91-99.

Gao, J., Chen, K., Xie, R. 2010a. In vivo tumor-targetedfluorescence imaging using near-infrared non-cadmium quantum dots. *Bioconjug Chemis.*, **21**(4), 604-609.

Gao, J., Chen, X., Cheng, Z. 2010b. Near-Infrared Quantum Dots as Optical Probes for Tumor Imaging. *Curr Top Med Chem.*, **10**(12), 1147-1157.

Gao, X., Cui, Y., Levenson, R.M., Chung, L.W.K., Nie, S. 2004. In vivo cancer targeting and imaging with semiconductor quantum dots. *Nat Biotechnol.*, **22**(8), 969-976.

Gaponik, N., Talapin, D.V., Rogach, A.L., Hoppe, K., Shevchenko, E.V., Kornowski, A., Eychmuller, A., Weller, H. 2002. Thiol-capping of CdTe nanocrystals: an alternative to organometallic synthetic routes. . *J Phys Chem B*, **106**(29), 7177-7185.

Grein, F., Ramos, A.R., Venceslau, S.S., Pereira, I.A. 2013. Unifying concepts in anaerobic respiration: Insights from dissimilatory sulfur metabolism. *Biochim Biophys Acta.*, **1872**(2), 145-160.

Gurunathan, S., Kalishwaralal, K., Vaidyanathan, R., Deepak, V., Pandian, S.R.K., Muniyandi, J. 2009. Biosynthesis, purification and characterization of silver nanoparticles using *Escherichia coli*. *Colloids Surf. B*, **74**, 328-35.

Hamada, Y., Gonda, K., Takeda, M., Sato, A., Watanabe, M., Yambe, T., Satomi, S., Ohuchi, N. 2011. In vivo imaging of the molecular distribution of the VEGF receptor during angiogenesis in a mouse model of ischemia. *Blood*, **118**, 93-100.

Harikrishnan, H., Shine, K., Ponmurugan, K., Moorthy, I.G., Kumar, R.S. 2014. In vitro eco-friendly synthesis of cadmium sulfide nanoparticles using heterotrophic *Bacillus cereus*. *J Optoelectronic and Biomedical Materials.*, **6**(1), 1-7.

Hasegawa, U., Nomura, S.M., Kaul, S.C. 2005. Nanogel-quantum dot hybrid nanoparticles for live cell imaging. *Biochem Biophys Res Commun.*, **331**(4), 917-921.

Herbel, J.M., Blum, S.J., Oremland, S.R. 2003. Reduction of Elemental Selenium to Selenide: Experiments with Anoxic Sediments and Bacteria that Respire Se-Oxyanions. *Geomicrobiol J.*, **20**, 587-602.

Hodlur, M.R., Rabinal, K.M. 2014. A new selenium precursor for the aqueous synthesis of luminescent CdSe quantum dots. *Chem Eng J.*, **244**, 82-88.

Holmes, D.J., Richardson, J.D., Saed, S., Evans-Gowing, R., Russell, A.D., Sodeau, R.J. 1997. Cadmium-specific formation of metal sulfide 'Q-particles' by *Klebsiella pneumoniae*. *Microbiology*, **143**, 2521-2530

Hoshino, A., Fujioka, K., Oku, T., Suga, M., Sasaki, Y.F., Ohta, T., Yasuhara, M., Suzuki, K., Yamamoto, K. 2004. Physicochemical properties and cellular toxicity of nanocrystal quantum dots depend on their surface modification. *Nano Lett.*, **4**(11), 2163-2169.

Hosseini, M., Sarvi, N.M. 2015. Recent achievements in the microbial synthesis of semiconductor metal sulfide nanoparticles. *Mater Sci Semicond Process.*, **40**, 293-301.

Hosseini, R.M., Schaffie, M., Pazouki, M., Schippers, A., Ranjba, M. 2013. A novel electrically enhanced biosynthesis of copper sulfide Nanoparticles. *Mater Sci Semicond Process.*, **16**, 250-255.

Hyun, R.B., Chen, Y.H., Rey, A.D., Wise, F.W., Batt, C.A. 2007. Near-infrared fluorescence imaging with water-soluble lead salt quantum dots. *J Phys Chem. B*, **111**(20), 5726-5730.

Jacob, M.J., Lens, P.N.L., Mohan, B.R. 2016. Microbial synthesis of chalcogenide semiconductor nanoparticles: a review. *Microb Biotechnol.*, **9**(1), 11-21.

Jacob, M.J., Mohan, B.R., Bhat, K.U. 2014. Biosynthesis of lead selenide quantum rods in marine *Aspergillus terreus*. *Mater Lett.*, **124**, 279-281.

Jaiswal, J.K., Simon, S.M. 2004. Potentials and pitfalls of fluorescent quantum dots for biological imaging. *Trends Cell Biol.*, **14**, 497-504.

Jaiswal, K.J., Mattoussi, H., Mauro, M.J., Simon, M.S. 2003. Long-term multiple color imaging of live cells using quantum dot bioconjugates. *Nat Biotechnol.*, **21**(1), 47-51.

Jamieson, T., Bakhshi, R., Petrova, D., Pocock, R., Imani, M., Seifalian, A.M. 2007. Biological applications of quantum dots. *Biomaterials.*, **28**(31), 4717-4132.

Jeyadevi, R., Sivasudha, T., Ramesh, A.K., Ananth, A.D., Aseervatham, G.S., Kumaresan, K., Kumar, D.L., Jagadeeswari, S., Renganathan, R. 2013. Enhancement of anti arthritic effect of quercetin using thioglycolic acid-capped cadmium telluride quantum dots as nanocarrier in adjuvant induced arthritic Wistar rats. *Colloids Surf B.*, **112**, 255-263.

Jie, G.F., Wang, L., Zhang, S.S. 2011. Magnetic electrochemi-luminescent Fe$_3$O$_4$=CdSe–CdS nanoparticle/polyelectrolyte nanocomposite for highly efficient immunosensing of a cancer biomarker. *Chem Eur J.*, **17**, 64-648.

Johns, T.A. 1951. Cytochrome b and bacterial succinic dehydrogenase. *Biochem. J.*, **49**, 559-560.

Jong, T., Parry, D.L. 2003. Removal of sulfate and heavy metals by sulfate reducing bacteria in short-term bench scale upflow anaerobic packed bed reactor runs. *Water res.*, **37**(14), 3379-3389.

Joo, J., Na, H.B., Yu, T., Yu, H.J., Hyeon, T. 2003. Generalized and facile synthesis of semiconducting metal sulfide nanocrystals. *J Am Chem Soc.*, **125**(36), 11100-11105.

Kairdolf, B.A., Smith, A.M., Stokes, T.H. 2013. Semiconductor quantum dots for bioimaging and biodiagnostic applications. *Annu Rev Anal Chem.*, **6**, 143-162.

Kaksonen, A.H., Riekkola-Vanhanen, M.L., Puhakka, J.A. 2003. Optimization of metal sulphide precipitation in fluidized-bed treatment of acidic wastewater. *Water res.*, **37**(2), 255-266.

Kim, S., Lim, Y.T., Soltesz, E.G., de Grand, A.M., Lee, J., Nakayama, A., Mihaljevic, T., Frangioni, J.V. 2004. Near-infrared fluorescent type II quantum dots for sentinel lymph node mapping. *Nat Biotechnol.*, **22**(1), 93-97.

Kirchner, C., Liedl, T., Kudera, S., Pellegrino, T., Muñoz Javier, A., Gaub, H.E., Stölzle, S., Fertig, N., Parak, W.J. 2005. Cytotoxicity of colloidal CdSe and CdSe/ZnS nanoparticles. *Nano Lett.*, **5**(2), 331-338.

Klonowska, A., Heulin, T., Vermeglio, A. 2005. Selenite and tellurite reduction by Shewanella oneidensis., *Appl Environ Microbiol.*(71), 9.

Kobayashi, H., Hama, Y., Koyama, Y. 2007. Simultaneous multicolor imaging of five different lymphatic basins using quantum dots. *Nano Lett.*, **7**, 1711-1716.

Kowshik, M., Deshmukh, N., Vogel, W., Urban, J., Kulkarni, K.S., Paknikar, K.M. 2002a. Microbial synthesis of semiconductor CdS nanoparticles, their characterization, and their use in the fabrication of an ideal diode. *Biotechnol Bioeng.*, **78**(5), 583-588.

Kowshik, M., Vogel, W., Urban, J., Kulkarni, K.S., Paknikar, K.M. 2002b. Microbial Synthesis of Semiconductor PbS Nanocrystallites. *Adv Mater.*, **14**(11), 815-818.

Krumov, N., Oder, S., Perner-Nochta, I., Angelov, A., Posten, C. 2007. Accumulation of CdS nanoparticles by yeasts in a fed-batch bioprocess. *J Biotechnol.*, **132**, 481-486.

Kumar, S.A., Ansary, A.A., Ahmad, A., Khan, M.I. 2007. Extracellular biosynthesis of CdSe Quantum Dots by the fungus, *Fusarium Oxysporum*. *J Biomed Nanotechnol.*, **3**(2), 190-194.

Labrenz, M., Druschel, G.K., Thomsen-Ebert, T., Gilbert, B., Welch, S.A., Kelly, S.D., Banfield, J.F. 2000. Formation of sphalerite (ZnS) deposits in natural biofilms of sulfate-reducing bacteria. *Science*, **290**(5497), 1744-1747.

Lenz, M., Lens, P.N.L. 2009. The essential toxin: The changing perception of selenium in environmental sciences. *Sci Total Environ.*, **407**(12), 3620-3633.

Li, D.B., Cheng, Y.Y., Wu, C., Li, W.W., Li, N., Yang, Z.C., Tong, Z.H., Yu, H.Q. 2014. Selenite reduction by *Shewanella oneidensis* MR-1 is mediated by fumarate reductase in periplasm. *Sci Rep.*, **4**, 3735.

Li, H., Shih, W.Y., Shih, W.H. 2007. Synthesis and characterization of aqueous carboxyl-capped CdS quantum dots for bioapplications. *Ind Eng Chem Res.*, **46**(7), 2013-2019.

Li, L., Wanga, L., Hub, T., Zhanga, W., Chen, X. 2014. Preparation of highly photocatalytic active CdS/TiO2 nanocomposites by combining chemical bath deposition and microwave-assisted hydrothermal synthesis. *J Solid State Chem.*, **218**, 81-89.

Li, X., Deng, D., Xue, J., Qu, L., Achilefu, S., Gu, Y. 2014. Quantum dots based molecular beacons for in vitro and in vivo detection of MMP-2 on tumor. *Biosens Bioelectron.*, **15**(61), 512-518.

Lin, D., Ji, J., Long, Z., Yang, K., Wu, F. 2012. The influence of dissolved and surface-bound humic acid on the toxicity of TiO2 nanoparticles to *Chlorella* sp. *Water Res.*, **46**, 4477-4487.

Liu, J., Hurt, R.H. 2010. Ion release kinetics and particle persistence in aqueous nano-silver colloids. *Environ Sci Technol.*, **44**(6), 2169-2175.

Liu, P., Wang, Q.S., Li, X. 2009. Studies on CdSe/L-cysteine quantum dots synthesized in aqueous solution for biological labeling. *J Phys Chem B*, **11**(8), 7670-7676.

Lloyd, J.R., Byrne, J.M., Coker, V.S. 2011. Biotechnological synthesis of functional nanomaterials. *Curr Opin Biotechnol.*, **22**(4), 509-515.

Lovrić, J., Bazzi, H.S., Cuie, Y., Fortin, G.R., Winnik, F.M., Maysinger, D. 2005. Differences in subcellular distribution and toxicity of green and red emitting CdTe quantum dots. *J Mol Med (Berl).* **83**(5), 377-385.

Luoa, X., Guoa, B., Wanga, L., Denga, F., Au, C. 2014. Synthesis of magnetic ion-imprinted fluorescent CdTe quantum dots by chemical etching and their visualization application for selective removal of Cd (II) from water. *Colloids Surf. A*, **462**, 186-193.

Maa, J., Wua, L., Houa, Z., Songc, Y., Wanga, L., Jiangb, W. 2014. Visualizing the endocytosis of phenylephrine in living cells by quantum dot-based tracking. *Biomaterials*, **35**(25), 7042-7049.

Mal, J., Nancharaiah, Y.V., van Hullebusch, E.D., Lens, P.N.L. 2016. Effect of heavy metal co-contaminants on selenite bioreduction by anaerobic granular sludge. *Bioresour Technol.*, **206**, 1-8.

Malarkodi, C., Annadurai, G. 2013. A novel biological approach on extracellular synthesis and characterization of semiconductor zinc sulfide nanoparticles. *Appl Nanosci.*, **3**, 389-395.

McConnachie, A.L., Botta, D., White, C.C., Weldy, C.S., Yu, J., Dills, R., Yu, X., Kavanagh, J.T. 2013. The Glutathione Synthesis Gene Gclm Modulates Amphiphilic Polymer-Coated CdSe/ZnS Quantum Dot–Induced Lung Inflammation in Mice. *PLoS One.*, **8**(5), 64165.

Medintz, I.L., Uyeda, H.T., Goldman, E.R., Mattoussi, H. 2005. Quantum dot bioconjugates for imaging, labelling and sensing. *Nat Mater.*, **4**(6), 435-446.

Mehra, R., Winge, D. 1991. Metal ion resistance in fungi: molecular mechanisms and their regulated expression. *J Cell Biochem.*, **43**, 30-40.

Meixner, M., Schöll, E., Shchukin, V.A., Bimberg, D. 2001. Self-assembled quantum dots: crossover from kinetically controlled to thermodynamically limited growth. *Phys Rev Lett.*, **87**(23), 236101.

Mi, C., Wang, Y., Zhang, J., Huang, H., Xu, L., Wang, S., Fang, X., Xu, S. 2011. Biosynthesis and characterization of CdS quantum dots in genetically engineered *Escherichia coli. J Biotechnol.*, **153**, 125-132.

Michalet, X., Pinaud, F.F., Bentolila, L.A. 2005. Quantum Dots for live cells, in vivo imaging, and diagnostics. *Science*, **307**(5709), 538-544.

Monras, P.J., Diaz, V., Bravo, D., Montes, A.R., Chasteen, G.T. 2012. Enhanced Glutathione Content Allows the In Vivo Synthesis of Fluorescent CdTe Nanoparticles by *Escherichia coli. PLoS ONE*, **7**(11), 48657.

Moon, J.W., Ivanov, N.I., Joshi, C.P., Armstron, L.B. 2014. Scalable production of microbially mediated zinc sulfide nanoparticles and application to functional thin films. *Acta Biomater.*, **10**, 4474-4483.

Moon, J.W., Roh, Y., Lauf, R.J., Vali, H., Yeary, L.W., Phelps, T.J. 2007. Microbial preparation of metal-substituted magnetite nanoparticles. *J Microbiol Method.*, **10**, 8298-8306.

Moore, M.D., Kaplan, S. 1994. Members of the family *Rhodospirilla ceaereduce* heavy-metal oxyanions to maintain redox poise during photosynthetic growth. *ASM News*, **60**, 17-24.

Moreau, J.W., Weber, P.K., Martin, M.C., Gilbert, B., Hutcheon, I.D., Banfield, J.F. 2007. Extracellular proteins limit the dispersal of biogenic nanoparticles. *Science.*, **316**, 1600-1603.

Mubarakali, D., Gopinath, V., Rameshbabu, N., Thajuddin, N. 2012. Synthesis and characterization of CdS nanoparticles using C-phycoerythrin from the marine cyanobacteria. *Mater Lett.*, **74**, 8-11.

Murray, C.B., Noms, D.J., Bawendi, M.G. 1993. Synthesis and characterization of nearly monodisperse CdE (E ¼ S, Se, Te) semiconductor nanocrystallites. *J Am Chem Soc.*, **115**, 8706-8715.

Mussa, F.S., Valizadeh, A. 2012. Review: three synthesis methods of CdX (X = Se, S or Te) quantum dots. *IET Nanobiotechnol.*, **8**(2), 59-76.

Muyzer, G., Stams, A.J. 2008. The ecology and biotechnology of sulphate-reducing bacteria. *Nat Rev Microbiol.*, **6**(6), 441-454.

Nancharaiah, Y.V., Lens, P.N. 2015. Ecology and biotechnology of selenium-respiring bacteria. *Microbiol Mol Biol Rev.*, **79**(1), 61-80.

Nancharaiah, Y.V., Lens, P.N. 2015b. Selenium biomineralization for biotechnological applications. *Trends Biotechnol.*, **33**, 323-330.

Narayanan, K.B., Sakthivel, N. 2010. Biological synthesis of metal nanoparticles by microbes. *Adv Colloid Interface Sci.*, **156**, 1-13.

Oh, S.H., Lim, S.C. 2006. A rapid and transient ROS generation by cadmium triggers apoptosis via caspase-dependent pathway in HepG2 cells and this is inhibited through N-acetylcysteine-mediated catalase upregulation. *Toxicol Appl Pharmacol.*, **212**(3), 212-223.

Ohde, H., Ohde, M., Bailey, F., Kim, H., Wai, C.M. 2002. Water-in-CO_2 microemulsions as nanoreactors for synthesizing CdS and ZnS nanoparticles in supercritical CO_2. *Nano Lett.*, **2**, 721-724.

Pandian, S., Deepak, V., Kalishwaralal, K., Gurunathan, S. 2011. Biologically synthesized fluorescent CdS NPs encapsulated by PHB. *Enzyme Microb Technol.*, **48**, 319-325.

Park, J., Dvoracek, C., Lee, K.H. 2011. CuInSe/ZnS core/shell NIR quantum dots for biomedical imaging. *Small*, **7**(22), 3148-3152.

Pawar, V., Kumar, A.R., Zinjarde, S., Gosavi, S. 2013. Bioinspired inimitable cadmium telluride quantum dots for bioimaging purposes. *J Nanosci Nanotechnol.*, **13**(6), 3826-3831.

Perez-Donoso, J.M., Monra's, J.P., Bravo, D., Aguirre, A., Quest, A.F., Osorio-Roman, I.O., Aroca, R.F., Chasteen, T.G., Vasquez, C.C. 2012. Biomimetic, mild chemical synthesis of CdTe-GSH quantum dots with improved biocompatibility. *PLoS ONE*, **7**(1), 30741.

Pearce, C.I., Coker, V.S., Charnock, J.M., Pattrick, R.A.D., Mosselmans, J.F.W., Law, N., Lloyd, R. 2008. Microbial manufacture of chalcogenide-based nanoparticles via the reduction of selenite using *Veillonella atypical* an in situ EXAFS study. *Nanotechnology*, **19**(5), 156603-156615.

Pearce, C.I., Pattrick, R.A.D., Law, N., Charnock, J.M., Coker, V.S., Fellowes, W.J., Oremland, S.R., Lloyd, R.J. 2009. Investigating different mechanisms for biogenic selenite transformations: *Geobacter sulfurreducens*, *Shewanella oneidensis* and *Veillonella atypica*. *Environ Technol.*, **30**(12), 1313-1326.

Peltier, E., Ilipilla, P., Fowle, D. 2011. Structure and reactivity of zinc sulfide precipitates formed in the presence of sulfate-reducing bacteria. *Appl Geochem.*, **26**, 1673-1680.

Prasad, K., Jha, K.A. 2010. Biosynthesis of CdS nanoparticles: An improved green and rapid procedure. *J Colloid Interf Sci.*, **342**, 68-72.

Qia, P., Zhang, D., Wan, Y. 2013. Sulfate-reducing bacteria detection based on the photocatalytic property of microbial synthesized ZnS nanoparticles. *Anal Chim Acta.*, **800**, 65-70.

Rao, N.N., Kornberg, A. 1996. Inorganic polyphosphate supports resistance and survival of stationary-phase *Escherichia coli. J Bacteriol.*, **178**, 1394-1400.

Rasmussen, E.T., Jauffred, L., Brewer, J., Vogel, S., Torbensen, E., Lagerholm, C.B., Oddershede, L., Arnspang, C.E. 2013. Single Molecule Applications of Quantum Dots. *J Modern Physics.*, **4**(11B), 27-42.

Rogach, A.L. 2008. Semiconductor Nanocrystal Quantum Dots,. *Springer-Verlag Vienna.*

Rzigalinski, B.A., Strobl, J.S. 2009. Cadmium-containing nanoparticles: Perspectives on pharmacology and toxicology of quantum dots. *Toxicol Appl Pharmacol.*, **238**, 280-288.

Sabaty, M., Avazeri, C., Pignol, D., Vermeglio, A. 2001. Characterization of the reduction of selenate and tellurite by nitrate reductases. *Appl Environ Microbiol.*, **67**, 5122-5126.

Sandana, M.J.G., Rose, C. 2014. Facile production of ZnS quantum dot nanoparticles by *Saccharomyces cerevisiae* MTCC 2918. *J Biotechnol.*, **170**, 73-78.

Sanghi, R., Verma, P. 2009. A facile green extracellular biosynthesis of CdS nanoparticles by immobilized fungus. *Chem Eng J.*, **155**, 886-891.

Sapsford, K., Pons, T., Medintz, I., Mattoussi, H. 2006. Biosensing with luminescent semiconductor quantum dots. *Sensors*, **6**(8), 925-953.

Schröfel, A., Kratošová, G., Šafařík, I., Šafaříková, M., Raška, I., Shor, L.M. 2014. Applications of biosynthesized metallic nanoparticles – A review. *Acta Biomater.*, **10**(10), 4023-4042.

Seshadri, S., Saranya, K., Kowshik, M. 2011. Green Synthesis of Lead Sulfide Nanoparticles by the Lead Resistant Marine Yeast, *Rhodosporidium diobovatum*. *Biotechnol Prog.*, **27**(5), 1464-1469.

Sharma, P.K., Balkwill, D.L., Frenkel, A., Vairavamurthy, M.A. 2000. A New *Klebsiella planticola* Strain (Cd-1) Grows Anaerobically at High Cadmium Concentrations and Precipitates Cadmium Sulfide. *Appl Environ Microbiol.*, **66**(7), 3083-3087.

Silver, S. 1996. Bacterial resistances to toxic metal ions-a review. *Gene.*, **179**(1), 9-19.

Singh, B.J., Dwivedi, S., Al-Khedhairy, A., Musarrat, J. 2011. Synthesis of stable cadmium sulfide nanoparticles using surfactin produced by *Bacillus amyloliquifaciens* strain KSU-109. *Colloids Surf. B*, **85**, 207-213.

Smith, B.R., Cheng, Z., De, A., Koh, A.L., Sinclair, R., Gambhir, S.S. 2008. Real-time intravital imaging of RGD-quantum dot binding to luminal endothelium in mouse tumor neovasculature. *Nano Lett.*, **8**(9), 2599-2606.

Speiser, D.M., Ortiz, D.F., Kreppel, L., Scheel, G., McDonald, G., Ow, D.W. 1992. Purine biosynthetic genes are required for cadmium tolerance in *Schizosacchromyces pombe*. . *Mol Cell Biol.*, **12**, 5301-5310.

Srivastava, N., Mukhopadhyay, M. 2013. Biosynthesis and structural characterization of selenium nanoparticles mediated by *Zooglea ramigera*. *Powder Technol.*, **244**, 26-29.

Steinberg, N.A., Oremland, R.S. 1990. Dissimilatory selenate reduction potentials in a diversity of sediment types. *Appl Environ Microbiol.*, **56**, 3550-3557.

Stolz, J.F., Basu, P., Santini, J.M., Oremland, R.S. 2006. Arsenic and selenium in microbial metabolism. *Annu Rev Microbiol.*, **60**, 107-130.

Suresh, K.A. 2014. Extracellular bio-production and characterization of small monodispersed CdSe quantum dot nanocrystallites. *Spectrochim Acta Mol Biomol Spectrosc.*, **130**, 344-349.

Sweeney, Y.R., Mao, C., Gao, X., Burt, L.J., Belcher, M.A., Georgiou, G., Iverson, L.B. 2004. Bacterial Biosynthesis of Cadmium Sulfide Nanocrystals. *Chem Biology*, **11**, 1553-1559.

Switzer, B.J., Burns, B.A., Buzzelli, J., Stolz, J.F., Oremland, R.S. 1998. *Bacillus arsenicoselenatis* sp. nov., and *Bacillus selenitireducens* sp. nov. : two haloalkaliphiles from Mono Lake, California which respire oxyanions of selenium and arsenic. *Arch Microbiol.*, **171**, 19-30.

Syed, A., Ahmad, A. 2013. Extracellular biosynthesis of CdTe quantum dots by the fungus *Fusarium oxysporum* and their anti-bacterial activity. *Spectrochim Acta Mol Biomol Spectrosc.*, **106**, 41-47.

Talapin, D.V., Poznyak, S.K., Gaponik, N.P., Rogach, A.L., Eychmüller, A. 2002. Synthesis of surface- modified colloidal semiconductor nanocrystals and study of photoinduced charge separation and transport in nanocrystal-polymer composites. *Physica.*, **14**, 237-241.

Trindade, T., O'Brien, P., Pickett, N.L. 2001. Nanocrystalline semiconductors: synthesis, properties, and perspectives. *Chem Mater.*, **13**, 3843-3458.

Turner, R.J., Borghese, R., Zannoni, D. 2012. Microbial processing of tellurium as a tool in biotechnology. *Biotechnol Adv.*, **30**, 954-963.

Valizadeh, A., Mikaeili, H., Samiei, M., Farkhani, S.M., Zarghami, N., Kouhi, M., Akbarzadeh, A., Davaran, S. 2012. Quantum dots: synthesis, bioapplications, and toxicity. *Nanoscale Res Lett.*, **7**(1), 480-494.

Vibin, M., Vinayakan, R., John, A., Fernandez, B.F., Abraham, A. 2014. Effective cellular internalization of silica-coated CdSe quantum dots for high contrast cancer imaging and labelling applications. *Cancer Nanotechnol.*, **5**, 1.

Wachter, J. 2004. Metal telluride clusters - from small molecules to polyhedral structures. *Eur J Inorg Chem.*, 1367-1378.

Wada, Y., Kuramoto, H., Anand, T., Tikamura, T. 2001. Microwave-assisted size control of CdS nanocrystallites. *J Mater Chem.*, **11**, 1936-1940.

Walling, A.M., Novak, A.J., Shepard, R.E.J. 2009. Quantum Dots for Live Cell and In Vivo Imaging. *Int J Mol Sci.*, **10**(2), 441-491.

Wang, C.L., Maratukulam, P.D., Lum, A.M., Clark, D.S., Keasling, J.D. 2000. Metabolic engineering of an aerobic sulfate reduction pathway and its application to precipitation of cadmium on the cell surface. *Appl Environ Microbiol.*, **66**, 4497-4502.

Wang, L.W., Peng, C.W., Chen, C., Li, Y. 2015. Quantum dots-based tissue and in vivo imaging in breast cancer researches: current status and future perspectives. *Breast Cancer Res Treat.*

Wang, Y.L., Lu, J.P., Tong, Z.F. 2010. Rapid synthesis of CdSe nanocrystals in aqueous solution at room temperature. *Bull Mater Sci.*, **33**(5), 543.

Weng, C.K., Noble, C.O., Papahadjopoulos-Sternberg, B. 2008. Targeted tumor cell internalization and imaging of multifunctional quantum dot-conjugated immunoliposomes in vitro and in vivo. *Nano Lett.*, **8**(9), 2851-2857.

Weres, O., Jaouni, A.R., Tsao, L. 1989. The distribution, speciation and geochemical cycling of selenium in a sedimentary environment, Kesterson Reservoir, California, U.S.A. . *Appl Geochem.*, **4**, 543-563.

Wiecinski, P.N., Metz, K.M., Heiden, T.C., Louis, K.M., Mangham, A.N., Hamers, R.J., Heideman, W., Peterson, R.E., Pedersen, J.A. 2013. Toxicity of oxidatively degraded quantum dots to developing zebrafish (*Danio rerio*). *Environ Sci Technol.*, **47**(16), 9132-9139.

Williams, P., Keshavarz-Moore, E., Dunnill, P. 1996a. Efficient production of microbially synthesized cadmium sulfide quantum semiconductor crystallites *Enzyme Microb Technol.*, **19**, 206-213.

Williams, P., Keshavarz-Moore, E., Dunnill, P. 1996b. Production of cadmium sulphide microcrystallites in batch cultivation by *Schizosaccharomyces pombe*. *J Biotechnol.*, **48**, 259-267

Williams, P., Keshavarz-Moore, E., Dunnill, P. 2002. *Schizosaccharomyces pombe* fed-batch culture in the presence of cadmium for the production of cadmium sulphide quantum semiconductor dots. *Enzyme Microb Technol.*, **30**, 354-362.

Woolfolk, C.A., Whiteley, H.R. 1962. Reduction of inorganic compounds with molecular hydrogen by *Micrococcus lactilyticus*. . *J Bacteriol.*, **84**, 647-658.

Wu, X., Liu, H., Liu, J., Haley, K.N., Treadway, J.A., Larson, J.P. 2003. Immunofluorescent labeling of cancer marker Her2 and other cellular targets with semiconductor quantum dots. . *Nat Biotechnol.*, **21**, 41-46.

Xiao, Y., Barker, P.E. 2004. Semiconductor nanocrystal probes for human metaphase chromosomes. *Nucleic Acids Res.*, **32**(3), 28-33.

Xu, J., Teslaa, T., Wu, T.H., Chiou, P.Y., Teitell, M.A., Weiss, S. 2012. Nanoblade delivery and incorporation of quantum dot conjugates into tubulin networks in live cells. *Nano Lett.*, **12**, 5669-5672.

Xue, B., Deng, D.W., Cao, J., Liu, F., Li, X., Akers, W., Achilefu, S., Gu, Y.Q. 2012. Synthesis of NAC capped near infrared-emitting CdTeS alloyed quantum dots and application for in vivo early tumor imaging. *Dalton Trans.*, **41**(16), 4935-4947.

Yan, Z., Qian, J., Gu, Y., Su, Y., Ai, X., Wu, S. 2014. Green biosynthesis of biocompatible CdSe quantum dots in living *Escherichia coli* cells. *Mater Res Exp.*, **1**(1), 15401.

Yang, C., Zhou, X., Wang, L., Tian, X., Wang, Y., Pi, Z. 2009. Preparation and tunable photoluminescence of alloyed CdS_xSe_{1-x} nanorods. *J Mater Sci.*, **44**, 3015-3019.

Yang, F., Xu, Z., Wang, J., Zan, F., Dong, C., Ren, J. 2013. Microwave-assisted aqueous synthesis of new quaternary-alloyed CdSeTeS quantum dots; and their bioapplications in targeted imaging of cancer cells. *Luminescence*, **28**, 392-400.

Yildiz, I., McCaughan, B., Cruickshank, F.S., Callan, F.J., Raymo, F.M. 2009. Biocompatible CdSe-ZnS Core-shell quantum dots coated with hydrophilic polythiols. *Langmuir*, **25**(12), 7090-7096.

Yuan, Y., Zhang, J., An, L., Cao, Q., Deng, Y., Liang, G. 2014. Oligomeric nanoparticles functionalized with NIR-emitting CdTe/CdS QDs and folate for tumor-targeted imaging. *Biomaterials.*, **35**(27), 7881-7886.

Zannoni, D., Borsetti, F., Harrison, J.J., Turner, J.R. 2008. The bacterial response to the chalcogen metalloids Se and Te,. *Adv Microbial Physiol.* , **53**, 1-72.

Zehr, J.P., Oremland, R.S. 1987. Reduction of selenate to selenide by sulfate-reducing bacteria: experiments with cell suspensions and estuarine sediments. . *Appl Environ Microbiol.*, **53**, 1365-1369.

Zhang, B., Hou, W.Y., Ye, X.C., Fu, S.Q., Xie, Y. 2007. 1D tellurium nanostructures: photothermally assisted morphology-controlled synthesis and applications in preparing functional nanoscale materials. . *Adv Funct Mater.*, **17**, 486-492.

Zhang, H., Wang, L., Xiong, H., Hu, L., Yang, B., Li, W. 2003. Hydrothermal synthesis for high-quality CdTe nanocrystals. *Adv Mater.*, **15**(20), 1712-1715.

Zhang, M.Z., Yu, Y., Yu, R.N., Wan, M., Zhang, R.Y., Zhao, Y.D. 2013. Tracking the down-regulation of folate receptor-α in cancer cells through target specific delivery of quantum dots coupled with antisense oligonucleotide and targeted peptide. *Small.*, **9**, 4183-4193.

Zhang, W., Chen, Z., Liu, H., Zhang, L., Gao, P., Li, D. 2011. Biosynthesis and structural characteristics of selenium nanoparticles by *Pseudomonas alcaliphila*. *Colloids Surf. B*, **88**(1), 196-201.

Zheng, Y.G., Gao, S.J., Ying, J.Y. 2007. Synthesis and cell-imaging applications of glutathione-capped CdTe quantum dots. *Adv Mater.*, **19**, 376-380.

Zhou, Z., Bedwell, J.G., Li, R., Prevelige, P., Gupta, A. 2014. Formation mechanism of chalcogenide nanocrystals confined inside genetically engineered virus-like particles. *Sci Rep.*, **4**, 1-6.

Zhua, Y., Lia, C., Xua, Y., Wang, D. 2014. Ultrasonic-assisted synthesis of aqueous CdTe/CdS QDs in salt water bath heating. *J Alloys Compd.*, **608**, 141-147.

Zrazhevskiy, P., Gao, X. 2013. Quantum dot imaging platform for single-cell molecular profiling. *Nat Commun.*, **4**, 1619.

CHAPTER 3

Biological removal of selenate and ammonium by activated sludge in a sequencing batch reactor

This chapter has been published in modified form:

Mal, J., Nancharaiah, Y.V., van Hullebusch, E., Lens, P.N.L. 2017. Biological removal of selenate and ammonium by activated sludge in a sequencing batch reactor. *Bioresour Technol.* 229, 11-19

Chapter 3

Abstract

Wastewaters contaminated by both selenium and ammonium need to be treated prior to discharge into natural water bodies, but there are no studies on the simultaneous removal of selenium and ammonium. A sequencing batch reactor (SBR) was inoculated with activated sludge and operated for 90 days. The highest ammonium removal efficiency achieved was 98%, while the total nitrogen removal was 75%. Nearly a complete chemical oxygen demand removal efficiency was attained after 16 days of operation, whereas complete selenate removal was achieved only after 66 days. The highest total Se removal efficiency was 97%. Batch experiments showed that the total Se in the aqueous phase decreased by 21% with increasing initial ammonium concentration from 50 to 100 mg. L^{-1}. This study showed that SBR can remove both selenate and ammonium via, respectively, bioreduction and partial nitrification-denitrification and thus offer possibilities for treating selenium and ammonium contaminated effluents.

Keywords: Sequencing batch reactor, activated sludge, selenate bioreduction, elemental selenium, simultaneous nitrification and denitrification

3.1. Introduction

Many wastewaters such as mining effluents (acid mine drainage, coal mine wastewaters, uranium mine discharge and gold mine wastewaters), agricultural drainage and industrial effluents are contaminated by both selenium oxyanions and ammonium, and thus need to be treated before discharging into natural water bodies (Dale et al., 2015; Kapoor et al., 2003; Tan et al., 2016; Muscatello et al., 2008; Ríos et al., 2008; Uhrie et al., 1996; Zaitsev et al., 2008). Selenium in these wastewaters is mainly present as selenate and selenite at typical concentrations varying between 0.4 - 53 mg. L^{-1} and has become a matter of public and scientific attention (Mal et al., 2016a; Tan et al., 2016). While the current selenium discharge criteria for aquatic life and the proposed toxicity thresholds are debatable, selenium release and its pollution has become a serious concern in recent years (Tan et al., 2016). Physical and chemical treatment approaches such as adsorption, ion exchange, chemical reduction and reverse osmosis are available for selenium removal from contaminated waters. But microbial technologies are promising for the treatment of selenium-containing wastewaters, mainly due to low cost and robustness of the bioprocess (Nancharaiah & Lens, 2015a; Nancharaiah & Lens, 2015b).

The mining industry uses large amounts of explosives containing ammonium nitrate, which is among the main sources of ammonium in mining effluents (Papirio et al., 2014; Zaitsev et al., 2008). The ammonium concentration in mining wastewaters was reported to range from 2 to 83 mg NH_4^+. L^{-1} (Kapoor et al., 2003; Koren et al., 2000; Ríos et al., 2008; Zaitsev et al., 2008). Agricultural waste, domestic and industrial effluents can also contain large quantities of ammonium (Islam et al., 2009; Ríos et al., 2008). Ammonium and free ammonia can promote corrosion and eutrophication, reducing the oxygen content and increasing toxicity to living organisms in aquatic ecosystems (Paredes et al., 2007). Therefore, wastewaters containing both ammonium and selenium need to be treated before being released into the environment. Sequencing batch reactors (SBR) allow complete nitrogen removal by the simultaneous nitrification and denitrification (SND) process (Wang et al., 2015). The presence of high ammonium concentrations in wastewater can influence the selenium treatment process and vice-versa (Yang et al., 2009). Synchronous removal of ammonium (NH_4^+-N) and sulfate can be achieved by the co-existence of anaerobic ammonium oxidation (ANAMOX) bacteria with sulfate reducing bacteria (SRB) (Yang et al., 2009; Zhao et al.,

2006). However, to the best of our knowledge, simultaneous removal of ammonium and selenium has not yet been reported.

In biological treatment of selenium-containing wastewaters, a significant fraction of the biogenic elemental selenium nanoparticles (SeNPs) remains in the bioreactor effluents due to their colloidal nature (Buchs et al., 2013). This not only compromises the discharge criteria of the treatment process, but also raises the concern on the fate of discharged biogenic SeNPs in the environment in terms of their toxicity and potential re-oxidation to soluble forms (Buchs et al., 2013; Mal et al., 2016b). Recently, Jain et al. (2016) have shown that activated sludge has better entrapment abilities for the biogenic SeNPs formed from selenite reduction. The entrapment of biogenic SeNPs in turn improved the settleability and hydrophilicity of the activated sludge. Thus, the activated sludge process could offer a promising alternative treatment, wherein effective total selenium removal can be achieved without the need for a post-treatment step for removing biogenic SeNPs envisaged in the case of an anaerobic treatment system (Park et al., 2013).

There are only a few studies on the use of activated sludge for the treatment of selenium-rich wastewaters. Moreover, the effect of alternating aerobic - anoxic or anaerobic conditions on selenate bioreduction by and the fate of biogenic SeNPs in the activated sludge wastewater treatment system are unknown. The objective of this study was, therefore, to investigate the potential of simultaneous removal of NH_4^+ and Se(VI) using activated sludge. A SBR was chosen for the study as it allows maintaining alternating aerobic-anoxic conditions required for oxidation of NH_4^+ and reduction of selenate, respectively. The SBR was inoculated with activated sludge and fed with artificial mining effluent containing ammonium chloride and sodium selenate. Removal of ammonium-N and selenate was monitored during long term SBR operation. Batch experiments were conducted to investigate the effect of different concentrations of ammonium on selenate bioreduction.

3.2. Materials and Methods

3.2.1. Source of biomass and synthetic wastewater

Activated sludge was collected from a full-scale domestic waste water treatment plant in Harnaschpolder (The Netherlands; Jain et al., 2016) and used for inoculating the SBR. The reactor was inoculated with 400 mL of activated sludge (1.2 g mixed liquor volatile suspended solids (MLVSS) L^{-1}).

3.2.2.   Synthetic wastewater

The synthetic ammonium-selenate wastewater was prepared in deionized water and consisted of the following basic ingredients as described in Nancharaiah et al. (2008) (in g. L^{-1}): Na$_2$SeO$_4$, 1.6; NH$_4$Cl, 0.3; K$_2$HPO$_4$, 0.06; KH$_2$PO$_4$, 0.02; MgSO$_4$.7H$_2$O, 0.06; and CaCl$_2$.2H$_2$O, 0.03. Trace elements were provided by adding 0.1 mL of trace element mix per 1 L of simulated ammonium-selenate wastewater (Nancharaiah et al., 2008). The influent of the SBR was supplemented with 1.0 g L^{-1} sodium acetate as sole carbon source, corresponding to an organic loading rate (OLR) of ~1.8 g chemical oxygen demand (COD) L^{-1} d^{-1}. The pH of the simulated ammonium-selenate rich wastewater was 7.0 - 7.1.

3.2.3.   Sequencing batch reactor operation

A polyacrilic column with influent, effluent and aeration ports was set-up and operated with a 1.2 L working volume in SBR mode for three months at 30°C. The schematic of reactor configuration used for the experiments is shown in Fig. 3.1. It was fed with the simulated ammonium-selenate wastewater with 66% volume exchange ratio with three cycles per day.

The SBR operation was divided into two periods. During period - I, the SBR was operated for 60 days with an 8h cycle period consisting of the following steps: 3 h static fill (anoxic/anaerobic), 4 h aeration, 10 min settle, 5 min decant and 45 min idle period. In period – II, the SBR cycle period was altered to 4.5 h static fill (anoxic/anaerobic), 2.5 h mixing by aeration, 10 min settle, 5 min decant and 45 min idle periods. During the aeration phase, mixing was achieved by bubbled aeration at the reactor bottom using a porous stone connected to an aerator. The dissolved oxygen (DO) in the SBR during the aeration period was > 5 mg. L^{-1}.

Sampling ports of the SBR were named S1, S2 and S3 from the bottom to the top for studying one typical SBR cycle (Fig. 3.1.). Samples from the influent and effluent of the SBR were collected daily to monitor the overall performance of the SBR. Occasionally, samples were collected during the SBR cycle period at 1.5 and 3 h during the filling period and at 4, 5, 6 and 7 h during the aerobic period to determine removal profiles.

3.2.4. Batch experiments – effect of ammonium on selenium removal

Batch tests were conducted to investigate the effect of ammonium on selenate (0.25 mM or 20 mg Se. L^{-1}) bioreduction at different initial NH$_4^+$-N concentrations of 0, 50 and 100 mg. L^{-1} at pH 7.0. The tests were performed in 120 mL volume glass serum bottles. The serum bottles were inoculated with 10 mL activated sludge from the SBR after the 66th day in 90 mL simulated ammonium-selenate wastewater as described above. The bottles were incubated at 30°C on an orbital shaker set at 150 rpm for 36 h. All the batch experiments were performed either in duplicate or triplicate. Liquid samples were collected at regular time intervals for analyzing selenate (Se (VI)), total selenium, elemental selenium (Se(0)), ammonium (NH$_4^+$), nitrite (NO$_2^-$) and nitrate (NO$_3^-$). Suitable controls were kept by incubating bottles containing only mineral medium and ammonium chloride with sodium selenate, but without the sludge biomass.

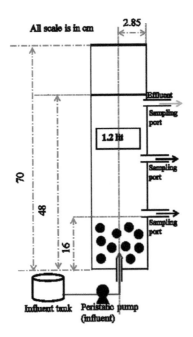

Fig. 3.1 Configuration of the sequencing batch reactor used for studying selenate and ammonium removal

3.2.5. Analytical methods

Liquid samples collected from the SBR were analyzed for NH_4^+, NO_2^-, NO_3^-, Se(VI), Se(0) and total selenium. Ammonium and nitrite were measured using the standard colorimetric methods as recommended by the US Environmental Protection Agency (EPA) (APHA, 2005). Nitrate and selenate were analyzed by an ion chromatograph (ICS-1000, IC, Dionex) equipped with a column (AS4A 2mm, Dionex) at retention times of, respectively, approximately 3.9 and 8.0 min. A 1.8 mmol Na_2CO_3 + 1.7 mmol $NaHCO_3$ solution was used as mobile phase at a flow rate of 0.5 mL/min. The chemical oxygen demand (COD) measurements were carried out as per standard methods (APHA, 2005).

The Se(0) and total selenium concentration were determined as described previously by Mal et al. (2016b). Briefly, Se(0) was collected from the liquid phase by centrifuging the samples at 37000 g for 15 min and the pellet was re-suspended in Milli-Q water. Total Se (before centrifugation) in the effluent and the Se(0) concentration in the pellet was determined using a graphite furnace AAS (SOLAAR MQZe, unity lab services USA) after acidifying with concentrated nitric acid (pH<2).

3.3. Results

3.3.1. Performance of the SBR

3.3.1.1. COD removal performance of the SBR

The COD removal profile of the SBR as a function of time is shown in Fig. 3.2. During the start-up, the SBR reactor performance was unstable and the COD removal efficiency was ~ 71%, while acetate utilization was ~75%. After 10 days of operation, the removal performance of the SBR became stable and the COD removal efficiency started to increase steadily to > 90% during days 10 - 18 (Fig. 3.2). Complete COD removal and acetate utilization was observed after 20 days of SBR operation, which remained the same until the end of period I. Alteration in the length of the aerobic and anaerobic phases of the SBR cycle period to 4.5 and 2.5 h in period II (days 60 – 90), respectively, had no effect on the COD removal efficiency or acetate utilization (Fig. 3.2): the COD removal efficiency and acetate utilization remained 100% until the end of the experiment.

Chapter 3

3.3.1.2.   Selenate removal performance of the SBR

The influent and effluent concentrations of Se(VI), Se(0) and total Se of the SBR are shown in Fig. 3.3. In period I, the SBR was operated with anaerobic and aerobic phases of 3 and 4 h, respectively. At the beginning of period I, the reactor performance was not stable and the selenate (1.6 mg. L^{-1}) removal efficiency was close to 70% (Fig. 3.3A). After two weeks of start-up, selenate reduction became stable and the removal efficiency increased steadily. The maximum total Se removal efficiency was 89.5% during period I. Selenate was not completely removed during the SBR cycle in period I and ~0.05 mg. L^{-1} selenate remained in the effluent (detection limit 0.05 mg. L^{-1}) until 60 d of reactor operation (Fig. 3.3A). Selenium speciation analysis, however, revealed that not selenate, but the elemental Se was the main contributor of the total Se in the effluent and its concentration ranged from 0.18 - 0.12 mg Se. L^{-1} during 30 to 60 days of SBR operation (Fig. 3.3B).

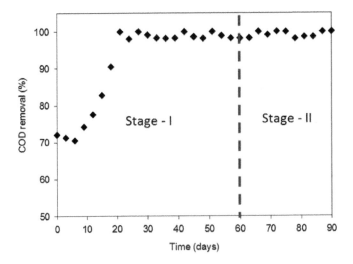

Fig. 3.2 COD removal efficiency of the activated sludge sequencing batch reactor. Period -I and Period -II refers to cycles with 3 h anaerobic - 4 h aerobic and 4.5 h anaerobic - 2.5 aerobic phases, respectively

Therefore, the aerobic and anaerobic phases of the SBR cycle period were altered to 4.5 and 2.5 h in period II, respectively, which lead to improvement of the Se(VI) removal efficiency during period II of SBR operation: a Se(VI) removal efficiency close to 100% was observed from day 66 of reactor operation onwards (Fig. 3.3B). Interestingly, the Se(0) concentration in the effluent of the SBR also decreased during period II and became as low as 0.043 mg Se. L^{-1} on day 75 of the reactor operation (Fig. 3.3B). As a result, the total Se in the effluent also decreased and the total Se removal efficiency reached to above 96% towards the end of the operation (day 90). The concentration of total Se in the treated waters was in the range of 0.05 to 0.06 mg Se L^{-1} (Fig. 3.3B), which is close to the US EPA limit of selenium (0.05 mg. L^{-1}), but higher than the flue gas desulfurization wastewater discharge criterion of 0.005 mg. L^{-1} (US EPA, 2015; US EPA Drinking water contaminants)

3.3.1.3.   Ammonium-N removal in the SBR

NH_4^+ removal was very slow during the start-up period (Fig. 3.3). From days 3 to 21, the effluent ammonium concentrations decreased with time from 44 to 24 mg. L^{-1} and the NH_4^+ removal efficiency reached up to 50% (Fig. 3.3C). After that, the NH_4^+ removal efficiency was continuously improved. After 48 days of operation, the NH_4^+ removal efficiency reached a maximum of $\sim$ 97% and was stable until the 60th day (end of period I) of SBR operation (Fig. 3.3D). However, the total nitrogen removal efficiency was only 52.2% due to incomplete denitrification (Fig. 3.3C). NO_3^- concentrations in the effluent were low during days 1 to 18 and increased to ~23 mg. L^{-1} on day 51. This was due to progressive establishment of nitrification coupled to incomplete denitrification, thus leading to the accumulation of residual nitrate in the reactor (Fig. 3.3C).

Nitrification slightly decreased immediately after the start of period II of SBR operation. When the aeration phase of the SBR cycle was decreased from 4 to 2.5 h, the effluent NH_4^+ concentrations increased marginally to 3.6 mg. L^{-1}, but slowly improved again. The effluent NH_4^+ concentration was $\leq$ 2.50 mg. L^{-1} until 90 days of SBR operation (Fig. 3.3C). Fig. 3.3C shows that though there was a marginal decrease in ammonium removal efficiency (from 97 to ~95%) during period II, the total N removal efficiency increased markedly from 50% to 75% because of decreasing NO_3^- concentrations in the effluent from days 60 onwards up to ~10 mg. L^{-1} of NO_3^- on the 90th day of the operation.

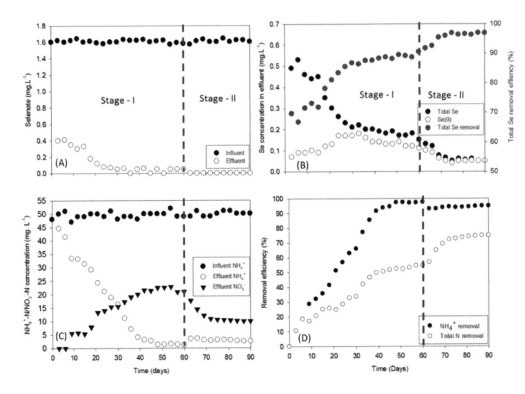

Fig. 3.3 Evolution of the SBR performance. (A-B) Selenate removal profiles and (C-D) Nitrogen removal profile during 90 days of reactor operation. Period - I: SBR cycle with 3 h anaerobic - 4 h aerobic phases. Period - II: SBR cycle with 4.5 h anaerobic - 2.5 h aerobic phases

3.3.2.   Simultaneous nitrogen and selenium removal profiles in SBR cycles

Time course profiles of Se concentrations during an SBR cycle on day 30, 57 and 87 are given in Fig. 3.4. During a typical cycle, the Se(VI) concentration increased steadily during the filling period and reached to its maximum at the end of the filling period. It is clear from Fig. 3.4 that the majority of the initial Se(VI) reduction happened during the static anaerobic filling phase. Smaller amounts of selenate (~0.2 - 0.3 mg. L^{-1}) remained at the end of the filling period which gradually decreased during the aeration phase. Se(VI) reduction became faster with time and 70% of selenate was reduced during the anaerobic static filling phase after 57 days of reactor operation (end of period II). At the end of period II, with the increase in anaerobic static filling phase, most of the initial Se(VI) (>90%) was reduced during the anaerobic phase only, prior to the aeration phase.

Although the majority of the selenate was removed by reduction prior to the aeration phase, there was no steady increase in Se(0) concentration during the anoxic phase (Fig. 3.4). However, there was a sudden increase in Se(0) formation within 1 h of the aeration period, which coincided with the selenate removal under aerobic conditions. Se(0) concentrations at the end of the aeration phase were 0.34, 0.31 and 0.16 mg Se. L^{-1}, at the 30th, 57th and 87th day of operation, respectively (Fig. 3.4).

Fig. 3.4 also shows the time course of the N-compounds in the SBR cycle periods on 30th, 57th and 87th day of SBR operation. During a typical cycle, the ammonium concentration increased during the filling period and reached a maximum at the end of the filling period: NH_4^+ was then nitrified to nitrite and nitrate during the aerobic phase. Interestingly, the NH_4^+ concentration in the SBR kept decreasing even during the filling (anaerobic) phases (Fig. 3.4). At the end of the anaerobic phase, the NH_4^+ concentration was 44.5, 38.5 and ~20 mg. L^{-1} at the 30th, 57th, and 87th day, respectively (Fig. 3.4). Nitrification during the aerobic phase became faster. On the 30th day, 17.41 mg. L^{-1} of initial NH_4^+ remained in solution after one 8h SBR cycle, while more than 90% of the NH_4^+ was oxidized and only 2 mg. L^{-1} of NH_4^+ was left on the 57th day. After the change in SBR cycle with an aeration phase of only 2.5 h, nitrification was marginally affected and 3.4 mg. L^{-1} of NH_4^+ remained in the effluent at the end of the SBR cycle period.

Fig. 3.4 shows that simultaneous nitrification and denitrification occurred in the SBR. The NO_3^- concentration inside the SBR increased at the beginning of each SBR cycle after 1.5 h. The NO_3^- concentration was 10.4 and 16.6 mg. L^{-1} inside the reactor after 1.5 h at the 30th and 57th day, respectively (Fig. 3.4). Denitrification took place mainly during the anoxic static filling phase and most of the NO_3^- was removed by the end of the anoxic phase (Fig. 3.4). $NO_3^- - N$ started accumulating again during the aerobic phase due to nitrification and at the end of the SBR cycle it reached to 22.2, 29.3 and 15.9 mg. L^{-1} at day 30, 57 and 87, respectively (Fig. 3.4)

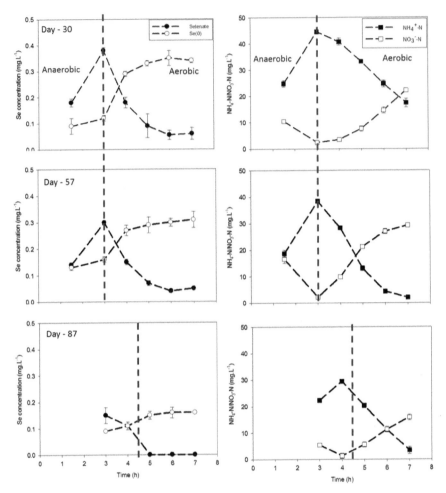

Fig. 3.4 Time course of N-compounds (ammonium-N, nitrate-N) (left) and Se-compounds (selenate and biogenic Se(0)) (right) during representative SBR cycles monitored on days 30, 57 and 87 of SBR operation. Day 30 and 57 belong to period-I and 87 belongs to periods-II of SBR operation. Period-I and period-II represent SBR cycle periods with 3 h anaerobic - 4 h aeration and 4.5 h anaerobic - 2.5 aeration phases, respectively. Influent selenate concentration was 1.6 mg. L^{-1}

3.3.3.   Effect of NH$_4^+$-N concentrations on selenium reduction

During batch tests with sludge sampled at the end of period I, Se(VI) concentrations were monitored as a function of time in the absence and presence of different concentration of NH$_4^+$

(50 and 100 mg. L^{-1}) in the anoxic phase (Fig. 3.5). There was no significant effect of NH_4^+ on Se(VI) bioreduction and Se(VI) was completely reduced at the two NH_4^+ concentrations tested. Complete Se(VI) reduction (initial Se(VI) concentration ~ 20 mg. L^{-1}) took around 30 h without or with 50 mg. L^{-1} NH$_4^+$-N (Fig. 3A). In the presence of high $NH_4^+ - N$ concentrations (100 mg. L^{-1}), Se(VI) bioreduction was marginally slower and complete within 36 h. Biomass associated Se increased by 14% with increasing initial NH$_4^+$-N concentration (Fig. 3.5B).

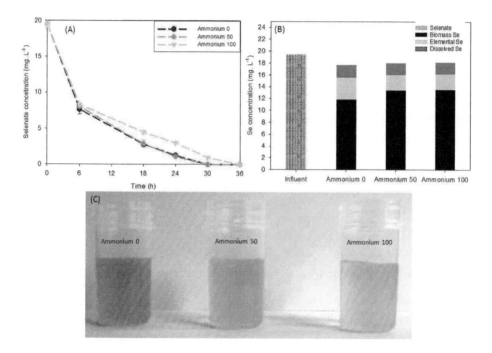

Fig. 3.5 Selenate removal profiles in batch experiments performed in serum bottles. A) Selenate removal profile at different initial concentrations of NH$_4^+$-N, B) Fate of selenium in the presence of different initial NH$_4^+$-N concentrations, C) Aqueous phase containing elemental Se after 36 h of incubation in the presence of different initial NH$_4^+$-N concentrations

The synthesis of Se(0) was clearly visible by the appearance of a orange-red color during the incubation period (Fig. 3.5). In the control experiment (without sludge biomass), there was neither selenate reduction nor a color change during the incubation period. Fig. 3.5B reveals the different Se(0) concentrations after the reduction of Se(VI) with and without $NH_4^+ - N$.

More amount of red elemental Se was produced in the absence of $NH_4^+ - N$ (3.72 mg. L^{-1}). In presence of $NH_4^+ - N$, there was no significant change in Se(0) concentration in the aqueous phase with the increase in initial NH_4^+-N concentration (Fig. 3.5B). However, total Se (elemental Se + dissolved Se) in the aqueous phase decreased by 21% with increasing initial NH_4^+-N concentration (Fig. 3.5B).

The time course of simultaneous NH_4^+ removal and other N-compounds during the anoxic phase is given in Fig. 3.6. It is clear that NH_4^+ removal occurred during the anoxic phase and it reduced to 40.08 and 85.23 mg. L^{-1} after 36 h of incubation in the presence of 50 and 100 mg. L^{-1} of initial NH_4^+ concentration, respectively (Fig. 3.6). No NO_3^- was observed during the incubation in either case. However, after 6 h of incubation, NO_2^- accumulation reached to 1.65 and 1.9 mg. L^{-1}, respectively, with 50 and 100 mg. L^{-1} NH_4^+ after 18 h of incubation. The NO_2^- concentration started to decrease afterwards and 0.95 mg. L^{-1} remained in the medium at the end of the incubation. In the presence of 100 mg. L^{-1} NH_4^+, NO_2^- removal was faster and only 0.42 mg. L^{-1} remained in the medium after 36 h of incubation. Unaccounted total N was 7.73 and 13.43 mg. L^{-1} in the presence of 50 and 100 mg. L^{-1} NH_4^+, respectively (Fig. 3.6)

3.4. Discussion

3.4.1. Selenate bioreduction by activated sludge in the presence of NH_4^+-N

This study shows for the first time that a selenate removal efficiency >95% can be achieved by activated sludge using the SBR configuration by alternating anaerobic and aerobic phases. Microbial reduction of both selenate and selenite under anaerobic conditions has been well documented (Nancharaiah & Lens, 2015a; Tan et al., 2016). But, there are limited to no studies on selenate reduction in aerobic reactors (Dhanjal & Cameotra, 2010; Staicu et al., 2015). Recently, activated sludge was used in a continuously operated reactor for studying selenite removal (Jain et al., 2016). Although selenite reduction was observed, the activated sludge reactor failed due to selenite toxicity to the microorganisms under aerobic conditions (Jain et al., 2016). In contrast, the present study demonstrated selenate removal efficiencies of >95% by the same activated sludge inoculum using the fill and draw SBR configuration with alternating anaerobic and aerobic phases in the cycle period (Fig. 3.3B). Stable selenate removal during long term reactor operation shows that the alternating anaerobic – aerobic phases inherent to SBR operation avoided selenium toxicity to the microorganisms present in the activated sludge system.

Selenate reduction by *Enterobacter cloacae* SLD1a-1, a facultative anaerobe was observed only after oxygen depletion from the nutrient medium (Losi & Frankenberger, 1997), indicating the influence of oxygen on selenate reduction. Yee et al. (2007) reported that selenate reduction by facultative anaerobes (selenate reductase gene) was regulated by oxygen-sensing proteins and occurs under suboxic conditions. Fig. 3.4 shows that most of the selenate was indeed removed during the anoxic static fill period. Although reduction of smaller amounts of selenate occurred under aerobic conditions, complete selenate removal was never achieved during period I of SBR operation (Fig. 3.3). The selenate removal efficiency of 70% on day 57 (end of period I) improved to > 90% on day 87 (end of period II) due to an increase in the length of the anoxic static fill period from 3 to 4.5 h (Fig. 3.4). Se(VI) removal was near complete with a 100% removal efficiency (Fig. 3.4) during period II indicating that selenate reducing microorganisms are more active under anoxic conditions than under aerobic conditions (Jain et al., 2016; Yee et al., 2007).

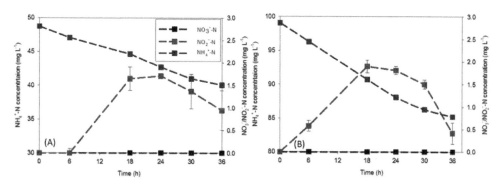

Fig. 3.6 NH_4^+-N, NO_3^--N and NO_2^--N profiles in batch experiments performed in serum bottles containing ammonium concentration of A) 50 mg. L^{-1}, and B) 100 mg. L^{-1}

Batch experiments confirmed that NH_4^+-N had no inhibitory effect on the microbial selenate reduction, even at 100 mg NH_4^+. L^{-1} (Fig. 3.5). Similarly, it was clearly evident that NH_4^+-N or NO_3^--N had no significant effect on the selenate reduction by activated sludge under anaerobic conditions, i.e. the filling period (Figs 3.5 and 3.6). This was not surprising, as distinct nitrate and selenate reductases are present in microorganisms like *Thauera selenatis* and *Sulfurospirillum barnesii* (Rech & Macy, 1992; Stolz et al., 1997). Simultaneous selenate reduction and denitrification under anaerobic conditions have been demonstrated previously in several studies (Dessì et al., 2016; Oremland et al., 1999; Rech & Macy, 1992). Simultaneous ammonium and sulfate removal has also been reported earlier (Yang et al., 2009; Zhao et al., 2006), but to the best of our knowledge this is the first study which shows the removal of both selenate and ammonium by activated sludge in SBR with alternating anoxic - aerobic phases in the cycle period.

Nitrate accumulation (produced due to nitrification of NH_4^+-N) could also negatively affect the selenate reduction by the activated sludge, particularly under aerobic conditions (Lai et al., 2014; Steinberg et al., 1992; Tan et al., 2016). The maximum nitrate accumulation was ~29 mg. L^{-1} in the SBR during period I, while it was only 15 mg. L^{-1} during period II (Fig. 3.3). Takada et al. (2008) reported that reduction of selenate to selenite by selenium-reducing sludge was not affected by the 500 mg. L^{-1} NO_3^- concentration, but selenite reduction to Se(0) was strongly inhibited even at 5 mg. L^{-1} NO_3^- concentration. Interestingly, no selenite accumulation was observed in the present study during aerobic conditions. So, it is possible that the lesser selenate reduction under aerobic conditions (Fig. 3.4) could be due to a combined effect of oxygen and nitrate accumulation (Takada et al., 2008; Yee et al., 2007).

3.4.2.   Fate of biogenic selenium in the SBR system

Selenate removal by the SBR was improved by providing a longer anaerobic phase from 3 to 4.5 h. This operational change has improved selenate removal efficiencies in the SBR. As a result, the selenate concentrations were below the detection limit in the effluent of the SBR from day 63 onwards (Fig. 3.3). Selenate removal was mainly mediated through microbial reduction as evident from the typical orange-red color of the elemental Se (Fig. 3.5). The Se(0) fraction in the treated water leaving the SBR was ~0.06 mg. L^{-1} Se accounted for <5% of the initial Se added (selenate) to the influent (Fig. 3.3). Se speciation analysis also showed that most of the total Se present in the effluent was in the form of Se(0) suggesting that the majority of selenate was reduced to Se(0) and not further to other forms of dissolved selenium, i.e. selenide (Se(-II)) or organo-Se. The batch experiments showed that the volatile Se fractions due to bio-methylation and bio-volatilization of selenate were also negligible (Fig. 3.5B). This has also been reported for Se treatment in other reactor configurations, like UASB (Lenz et al., 2008) and activated sludge (Jain et al., 2016; Jain et al., 2015) reactors, wherein the volatile Se fraction was negligible. The negligible presence of volatile selenium and the presence of elemental Se in low concentrations in the effluent suggest that most of the biogenic elemental Se formed was retained in the activated sludge flocs.

Another possible reason for the lower SeNPs concentration in the SBR effluent could be due the different characteristics, e.g. size, surface charge and colloidal stability of the Se nanoparticles (SeNPs) produced during the alternating anaerobic-aerobic conditions or denitrification in the SBR (Buchs et al., 2013; Jain et al., 2016). Previously, it was reported that nitrate influences the settling characteristics of activated sludge flocs and accelerates aerobic granulation due to denitrification (Suja et al., 2015). It is possible that the activated sludge growing in the SBR became compact and dense due to the denitrification process, which helped in higher entrapment of SeNPs and superior settling behavior of the activated sludge flocs.

It should be noted that the majority of the selenate was removed during the anoxic/anaerobic phase (static fill period) with traces of elemental Se in the liquid phase (Fig. 3.4). The minor amounts of selenate reduction during the aeration phase coincided with the release of elemental Se into the liquid phase. Reductive removal of selenate at the beginning of aeration phase is possible either due to the presence of aerobic selenate reducing microorganisms, such as *Pseudomonas stutzeri* and (Kuroda et al., 2011) and *Pseudomonas moraviensis* subsp. stanleyae (Staicu et al., 2015) or due to the occurrence of anoxic/anaerobic zones in the activated sludge where anaerobic bacteria can reduce the selenate via dissimilatory selenate reduction.

It is most likely that selenate reduction under aerobic conditions yielded relatively higher elemental Se(0) in the SBR (Fig. 3.4). This may be due to the operation of different microbial reduction mechanisms during the anoxic/anaerobic and aerobic phases. Another reason could be that elemental selenium is reduced further microbiologically to soluble selenide (Se(-II)) under anaerobic conditions (Mal et al., 2016b). Unlike other Se compounds, selenide can spontaneously and rapidly oxidize to elemental selenium in the presence of oxygen and thus increasing the Se(0) concentrations during aerobic condition compared to anaerobic condition (Nancharaiah & Lens, 2015).

The mechanisms of selenium oxyanion reduction and localization of biogenic Se(0) formation are not yet well established (Nancharaiah & Lens, 2015). Many bacteria use selenate as terminal electron acceptors and reduce soluble selenate to insoluble elemental selenium via dissimilatory reduction under anaerobic/anoxic conditions. In contrast, selenate is generally reduced to elemental selenium via detoxification mechanism under aerobic or microaerophilic conditions (Nancharaiah & Lens, 2015). Moreover, different reduction mechanisms (intracellular versus extracellular reduction) can occur in different microorganisms. It could be possible that under anoxic/aerobic conditions microorganisms follow different mechanisms for the selenate reduction, resulting in an increase of the intracellular Se(0) synthesis when selenate enters the cytoplasm via sulfate, nitrite, or an independent, as yet unidentified, transporter (Nancharaiah & Lens, 2015). Hapuarachchi et al. (2004), however, reported that sequential anaerobic/aerobic growth conditions did not increase the production of elemental selenium due to Se(IV) reduction by a pure culture of *Pseudomonas fluorescens* K27.

3.4.3. Simultaneous nitrification and denitrification

This study demonstrated ammonium-nitrogen was removed by biological nitrification-denitrification, using acetate as electron donor for the denitrification step. Although NH_4^+ removal was more than 95%, the total N removal efficiency was only ~75% due to the presence of NO_3^--N in the effluent (Fig. 3.3). The main reason of nitrate accumulation was the lack of external carbon source (i.e. acetate), because the majority of the COD was removed within 2 h during the aeration period. Thus, nitrate removal by heterotrophic denitrification was partially limited in the aerobic phase due to insufficient supply of electron donor (Jia et al., 2013; Suja et al., 2015).

The decrease in ammonium concentration even during the filling period could be due to the partial nitrification in the anoxic period (Fig. 3.4) (Breisha, 2010; Kornaros et al., 2010). Partial nitrification of ammonium-nitrogen can be achieved at low DO concentrations of 0.2 - 0.7 mg. L^{-1} (Wang et al., 2015). Such a low DO (< 0.5 mg. L^{-1}) concentration usually inhibits the activity of nitrite-oxidizing bacteria, but not the ammonia oxidizing bacteria resulting in the accumulation of nitrite (Hwang et al., 2009; Metcalf, 2013; Wang et al., 2015). The operating mode of an alternative anoxic and aerobic mode could induce nitrite accumulation and result in partial nitrification of ammonium up to nitrite and nitrite denitrification (Wang et al., 2015; Zhang et al., 2011). The increase in the length of the anoxic period during period II might foster the chance of partial nitrification and result in partial nitrification and denitrification, and thus higher TN removal efficiencies (Fig. 3.3).

This phenomenon was also observed in the batch experiments, wherein ammonium-nitrogen removal was associated with nitrite, but not nitrate, accumulation (Fig. 3.6). During partial nitrification and nitrite denitrification, ammonium is converted up to nitrite (via AOB) and nitrite is subsequently denitrified by denitrifying microorganisms to nitrogen (Kornaros et al., 2010). It is most likely that the low amounts of oxygen available in the medium and head space at the beginning of the incubation were consumed by the partial nitrification process (resulting in the accumulation of nitrite) by the activated sludge within the initial 18 h incubation (Fig. 3.6), thus making the conditions more suitable for denitrifying microorganisms. After 18 h of incubation, although the NH_4^+ concentration kept decreasing, the nitrite concentration also kept decreasing instead of increasing, indicating simultaneous ammonium oxidation to nitrite and denitrification of the produced nitrite.

Fig. 3.6 shows that higher initial ammonium concentrations were associated with higher ammonium-nitrogen removal efficiencies. After 36 h of incubation, 8.68 and 13.85 mg. L^{-1} of NH_4^+ was removed from the initial 50 and 100 mg. L^{-1} NH_4^+, respectively (Fig. 3.6). Similar trends have been reported previously when it was suggested that the nitrification rate increased with increasing ammonium loading rate (Hwang et al., 2009; Zhao et al., 2006). Interestingly, with the increase in the initial ammonium or selenate concentration, both the Se and TN removal efficiency increased during the batch experiments suggesting that further studies using e.g. stable isotopes of nitrogen (Xu et al., 2016) and selenium (Schilling et al., 2015) are required to determine the optimal NH_4^+ to Se(VI) ratio for simultaneous removal of ammonium and selenium to increase the treatment efficiencies of the SBR.

After increasing the length of the anoxic period, the total Se and TN removal efficiency increased, indicating the importance of the duration of both the anoxic or aerobic phase that need to be further optimized to further improve the nitrogen and selenium removal efficiencies. Multiple anoxic and aerobic periods could be another option to increase the SBR performance, where intermittent aeration can be adopted to achieve multiple nitrification-denitrification periods within one SBR cycle (Wang et al., 2015). Isolation of the functional bacteria and their subsequent morphological, biochemical and metabolic characterization is also desirable to have a better understanding of the effect of the oxygen concentration on the selenate bioreduction and localization of the elemental Se biosynthesis in the activated sludge microorganisms. Nonetheless, the results obtained here are useful for the development of a novel SBR based system capable of simultaneously removing ammonium and selenate from wastewaters.

3.5. Conclusion

Efficient removal of selenate and ammonium-nitrogen was demonstrated for the first time by activated sludge in SBR operated with alternating anaerobic-aerobic phases in the cycle. Selenate removal efficiencies of up to ~100% were achieved through microbial reduction. Ammonium removal efficiencies of up to 95% were mediated through partial nitrification as well as nitrification-denitrifiction. The fill and draw configuration of SBR along with anaerobic-aerobic phases in the cycle allowed removal of both selenate and ammonium. The efficient treatment performance of an SBR offers possibilities of treating mining effluents and agricultural drainage contaminated with both selenium and ammonium.

References

APHA. 2005. Standard methods for examination of water and wastewater, 5th ed. American Public Health Association, Washington, DC, USA.

Breisha, G.Z. 2010. Bio-removal of nitrogen from wastewaters - A review. *Nature and Science*, **8**(12), 210-229.

Buchs, B., Evangelou, M.W., Winkel, L.H., Lenz, M. 2013. Colloidal properties of nanoparticular biogenic selenium govern environmental fate and bioremediation effectiveness. *Environ Sci Technol.*, **47**(5), 2401-2407.

Dale, C., Laliberte, M., Oliphant, D., Ekenberg, M. 2015. Wastewater treatment using MBBR in cold climates. Proceedings of Mine Water Solutions in Extreme Environments. ISBN: 978-0-9917905-7-9

Dessì, P., Jain, R., Singh, S., Seder-Colomina, M., van Hullebusch, E.D., Rene, E.R., Ahammad, S.Z., Lens, P.N.L. 2016. Effect of temperature on selenium removal from wastewater by UASB reactors. *Water Res.*, **94**, 146-154.

Dhanjal, S., Cameotra, S.S. 2010. Aerobic biogenesis of selenium nanospheres by *Bacillus cereus* isolated from coal mine soil. *Microb Cell Fact.*, **9**, 52.

Hapuarachchi, S., Swearingen, J., Chasteen, G.T. 2004. Determination of elemental and precipitated selenium production by a facultative anaerobe grown under sequential anaerobic/aerobic conditions. *Process Biochem.*, **39**, 1607-1613.

Hwang, J.H., Cicek, N., Oleszkiewicz, J. 2009. Effect of loading rate and oxygen supply on nitrification in a non-porous membrane biofilm reactor. *Water Res.*, **43**(13), 3301-3307

Islam, M., George, N., Zhu, J., Chowdhury, N. 2009. Impact of carbon to nitrogen ratio on nutrient removal in a liquid-solid circulating fluidized bed bioreactor (LSCFB). *Process Biochem.*, **44**(5), 578-583.

Jain, R., Matassa, S., Singh, S., van Hullebusch, E.D., Esposito, G., Lens, P.N.L. 2016. Reduction of selenite to elemental selenium nanoparticles by activated sludge. *Environ Sci Pollut Res Int.*, **23**(2), 1193-202.

Jain, R., Seder-colomina, M., Jordan, N., Dessi, P., Cosmidis, J., van Hullebusch, E.D., Weiss, S., Farges, F., Lens, P.N.L. 2015. Entrapped elemental selenium nanoparticles affect physicochemical properties of selenium fed activated sludge. *J Hazard Mater.*, **295**, 193-200.

Jia, W., Liang, S., Zhang, J., Ngo, H.H., Guo, W., Yan, Y., Zou, Y. 2013. Nitrous oxide emission in low-oxygen simultaneous nitrification and denitrification process: sources and mechanisms. *Bioresour Technol.*, **136**, 444-451.

Kapoor, A., Kuiper, A., Bedard, P., Gould W.D. 2003. Use of a rotating biological contactor for removal of ammonium from mining effluents. *Eur. J. Miner. Process. Environ. Prot.* 3(1), 88-100.

Karya, N.G.A., van der Steen, P.N., Lens, P.N.L. 2013. Photo-oxygenation to support nitrification in an algal-bacterial consortium treating artificial wastewater. *Bioresour Technol.*, **134**, 244-250.

Koren, D.W., Gould, W.D., Bedard, P. 2000. Biological removal of ammonia and nitrate from simulated mine and mill effluents. *Hydrometallurgy* 56(2), 127-144.

Kornaros, M., Dokianakis, S.N., Lyberatos, G. 2010. Partial nitrification/denitrification can be attributed to the slow response of nitrite oxidizing bacteria to periodic anoxic disturbances. *Environ. Sci. Technol.* 44(19), 7245-7253.

Kuroda, M., Notaguchi, E., Sato, A., Yoshioka, M., Hasegawa, A., Kagami, T., Narita, T., Yamashita, M., Sei, K., Soda, S., Ike, M. 2011. Characterization of Pseudomonas stutzeri NT-I capable of removing soluble selenium from the aqueous phase under aerobic conditions. *J. Biosci. Bioeng.* 112(3), 259-264.

Lai, C.Y., Yang, X., Tang, C.Y., Rittmann, B.E., Zhao, H.P. 2014. Nitrate shaped the selenate-reducing microbial community in a hydrogen-based biofilm reactor. *Environ Sci Technol.*, **48**, 3395-3402.

Lenz, M., van Hullebusch, E.D., Hommes, G., Corvini, P.F., Lens, P.N.L. 2008. Selenate removal in methanogenic and sulfate-reducing upflow anaerobic sludge bed reactors. *Water Res.*, **42**, 2184–2194.

Losi, M.E., Frankenberger, W.T. 1997. Reduction of selenium oxyanions by *Enterobacter cloacae* SLD1a-1: isolation and growth of the bacterium and its expulsion of selenium particles. . *Appl Environ Microbiol.*, **63**, 3079-3084.

Mal, J., Nancharaiah, Y., van Hullebusch, E., Lens, P.N.L. 2016a. Metal Chalcogenide quantum dots: biotechnological synthesis and applications. *RSC Adv.* 6, 41477-41495.

Mal, J., Nancharaiah, Y., van Hullebusch, E., Lens, P.N.L. 2016. Effect of heavy metal co-contaminants on selenite bioreduction by anaerobic granular sludge. *Bioresour Technol.*, **206**, 1-8.

Metcalf, E. 2013. Wastewater Engineering, Treatment and Reuse, fourth ed. McGraw-Hill, New York.

Muscatello, J.R., Belknap, A.M., Janz, D.M. 2008. Accumulation of selenium in aquatic systems downstream of a uranium mining operation in northern Saskatchewan, Canada. *Environ. Pollut.* 156(2), 387-393.

Nancharaiah, Y.V., Joshi, H.M., Hausner, M., Venugopalan, V.P. 2008. Bioaugmentation of aerobic microbial granules with *Pseudomonas putida* carrying TOL plasmid. *Chemosphere*, **71**, 30-35.

Nancharaiah, Y.V., Lens, P.N.L. 2015. Ecology and biotechnology of selenium-respiring bacteria. *Microbiol Mol Biol Rev.*, **79**(1), 61-80.

Nancharaiah, Y.V., Lens, P.N.L. 2015b. Selenium biomineralization for biotechnological applications. *Trends Biotechnol.*, **33**, 323-330.

Oremland, R.S., Blum, J.S., Bindi, A.B., Dowdle, P.R., Herbel, M., Stolz, J.F. 1999. Simultaneous Reduction of Nitrate and Selenate by Cell Suspensions of Selenium-Respiring Bacteria. *Appl Environ Microbiol.*, **65**(10), 4385-4392.

Papirio, S., Zou, G., Ylinen, A., Di Capua, F., Pirozzi, F., Puhakka, J.A. 2014. Effect of arsenic on nitrification of simulated mining water. *Bioresour Technol.*, **164**, 149-154.

Paredes, D., Kuschk, P., Mbwette, T.S.A., Stange, F., Müller, R.A., Köser, H. 2007. New aspects of microbial nitrogen transformations in the context of wastewater treatment – a review. *Eng Life Sci.*, **7**, 13-25.

Park, H.J., Kim, H.Y., Cha, S., Ahn, C.H., Roh, J., Park, S., Kim, S., Yoon, J. 2013. Removal characteristics of engineered nanoparticles by activated sludge. *Chemosphere.*, **92**(5), 524-528.

Rech, S.A., Macy, J.M. 1992. The terminal reductases for selenate and nitrate respiration in *Thauera selenatis* are two distinct enzymes. *J Bacteriol.*, **17**, 7316-7320.

Ríos, C.A., Williams, C.D., Roberts, C.L. 2008. Removal of heavy metals from acid mine drainage (AMD) using coal fly ash, natural clinker and synthetic zeolites. *J Hazard Mater.*, **156**(1-3), 23-35.

Schilling, K., Johnson, T.M., Dhillon, K.S., Mason, P.R. 2015. Fate of Selenium in Soils at a Seleniferous Site Recorded by High Precision Se Isotope Measurements. *Environ. Sci. Technol.* 49(16), 9690-9698.

Staicu, L.C., Ackerson, C.J., Cornelis, P., Ye, L., van Hullebusch, E.D., Lens, P.N.L., Pilon-Smits, E.A. 2015. *Pseudomonas moraviensis* subsp. stanleyae, a bacterial endophyte of hyperaccumulator Stanleya pinnata, is capable of efficient selenite reduction to elemental selenium under aerobic conditions. *J Appl Microbiol.*, **119**(2), 400-410.

Steinberg, N.A., Switzer Blum, J., Hochstein, L., Oremland, R.S. 1992. Nitrate is a preferred electron acceptor for growth of selenate-respiring bacteria. *Appl Environ Microbiol.*, **56**, 426-428.

Stolz, J.F., Gugliuzza, T., Switzer Blum, J., Oremland, R., Martinez Murillo, F. 1997. Differential cytochrome content and reductase activity in *Geospirillum barnesii* strain SES3. . *Arch. Microbiol.*, **167**, 1-5.

Suja, E., Nancharaiah, Y.V., Krishna Mohan, T.V., Venugopalan, V.P. 2015. Denitrification accelerates granular sludge formation in sequencing batch reactors. *Bioresour Technol.*, **196**, 28-34.

Takada, T., Hirata, M., Kokubu, S., Toorisaka, E., Ozaki, M., Hano, T. 2008. Kinetic study on biological reduction of selenium compounds. *Process Biochem.*, **43**, 1304-1307.

Tan, L.C., Nancharaiah, Y., van Hullebusch, E., Lens, P.N.L. 2016. Selenium: environmental significance, pollution, and biological treatment technologies. *Biotechnol Adv.*, **34**(5), 886-907.

Uhrie, J.L., Drever, J.I., Colberg, P.J.S., Nesbitt, C.C. 1996. In situ immobilization of heavy metals associated with uranium leach mines by bacterial sulfate reduction. *Hydrometallurgy.* 43, 231-239.

US EPA. 2015. Effluent limitations guidelines and standards for the steam electric power generating point source category. Fed. Regist. 80 (212), 67900.

(https://www.gpo.gov/fdsys/pkg/FR-2015-11-03/pdf/2015-25663.pdf.)

US EPA Drinking water contaminants. http://www.epa.gov/your-drinking-water/table-regulated-drinking-water-contaminants#Inorganic.

Wang, H., Guan, Y., Li, L., Wu, G. 2015. Characteristics of Biological Nitrogen Removal in a Multiple Anoxic and Aerobic Biological Nutrient Removal Process. *Biomed Res Int.*, **2015**, 8.

Xu, S., Kang, P., Sun Y. 2016. A stable isotope approach and its application for identifying nitrate source and transformation process in water. *Environ. Sci. Pollut. Res Int.* 23(2), 1133-1148.

Yang, Z., Zhou, S., Sun, Y. 2009. Start-up of simultaneous removal of ammonium and sulfate from an anaerobic ammonium oxidation (anammox) process in an anaerobic up-flow bioreactor. *J Hazard Mater.*, **169**(1-3), 113-118.

Yee, N., Ma, J., Dalia, A., Boonfueng, T., Kobayashi, D.Y. 2007. Se(VI) Reduction and the Precipitation of Se(0) by the Facultative Bacterium *Enterobacter cloacae* SLD1a-1 Are Regulated by FNR. *Appl Environ Microbiol.*, **73**(6), 1914-1920.

Zaitsev, G., Mettanen, T., Langwaldt, J. 2008. Removal of ammonium and nitrate from cold inorganic mine water by fixed-bed biofilm reactors. *Miner. Eng.*, **21**, 10-15.

Zhang, M., Lawlor, P.G., Wu, G., Lynch, B., Zhan, X. 2011. Partial nitrification and nutrient removal in intermittently aerated sequencing batch reactors treating separated digestate liquid after anaerobic digestion of pig manure. *Bioprocess Biosyst Eng.*, **34**(9), 1049-1056.

Zhao, Q.I., Li, W., You, S.J. 2006. Simultaneous removal of ammonium-nitrogen and sulphate from wastewaters with an anaerobic attached-growth bioreactor. *Water Sci Technol.*, **54**(8), 27-35.

CHAPTER 4

Effect of heavy metal co-contaminants on selenite bioreduction by anaerobic granular sludge

This chapter has been published in modified form:

Mal, J., Nancharaiah, Y.V., van Hullebusch, E., Lens, P.N.L. 2016. Effect of heavy metal co-contaminants on selenite bioreduction by anaerobic granular sludge. *Bioresour Technol.* 206, 1-8

Chapter 4

Abstract

This study investigated bioreduction of selenite by anaerobic granular sludge in the presence of heavy metals and analyzed the fate of the bioreduced selenium and the heavy metals. Selenite bioreduction was not significantly inhibited in the presence of Pb(II) and Zn(II). More than 92% of 79 mg. L^{-1} selenite was removed by bioreduction even in the presence of 150 mg. L^{-1} of Pb(II) or 400 mg. L^{-1} of Zn(II). In contrast, only 65-48% selenite was bioreduced in the presence of 150 to 400 mg. L^{-1} Cd(II). Formation of elemental selenium or selenide varied with heavy metal type and concentration. Notably, the majority of the bioreduced selenium (70-90% in the presence of Pb and Zn, 50-70% in the presence of Cd) and heavy metals (80-90% of Pb and Zn, 60-80% of Cd) was associated with the granular sludge. The results have implications in the treatment of selenium wastewaters and biogenesis of metal selenides.

Keywords: anaerobic granular sludge; biosorption; heavy metal removal; metal selenide; selenite bioreduction.

4.1. Introduction

Selenium and sulfur belong to the chalcogen group (periodic table group 16), and have similar chemical behavior. Often, selenium oxyanion contamination occurs concomitantly with sulfate and heavy metals in different waste streams such as acid mine drainage, acid seeps, and agricultural drainage (Table 4.1). There are limited studies on the microbial transformation of selenium oxyanions as compared to sulfur (Nancharaiah et al., 2015a). Bioreduction of selenium oxyanions, particularly selenite, to elemental selenium can be achieved using both aerobic and anaerobic microorganisms (Nancharaiah et al., 2015a). Thus, selenate (Se(VI)) and selenite (Se(IV)) reducing microorganisms could be potentially used for the bioremediation of selenium contaminated soils, sediments, industrial effluents, and agricultural drainage waters (Lenz et al., 2008; Nancharaiah et al., 2015a, 2015b). However, the use of this strategy for practical applications may have important limitations because the microbial reduction processes as well as the fate of bioreduced species can be affected by the presence of co-contaminants such as heavy metals.

Table 4.1 Overview of selenium and metal concentrations in different waste streams

Waste stream	Se	Cd	Zn	Pb	Reference
	(mg. L^{-1})	(mg. L^{-1})	(mg. L^{-1})	(mg. L^{-1})	
Acid mine drainage	1-53	2-44	517-5000	37-1240	Moreau et al. (2013)
	0.01-0.03	0.01	-	0.02	Matlock et al. (2002)
	0.02	0.024	43	0.05	Espana et al. (2006)
	6	0.1	20	-	
Flue gas de-sulfurization	0.015-162	0.005-81.9	0.01-5070	0.01-527	Meawad et al. (2010)

Heavy metals are toxic for microorganisms and cannot be biodegraded like organic pollutants. However, they can be transformed from mobile and toxic forms into immobile and less or non-toxic forms (Beyenal et al., 2004; Nancharaiah et al., 2015c). Both adsorption and redox conditions essentially control the mobility of these chemical species in natural environments. It has been well documented that heavy metals such as Cu(II), Zn(II) and Cd(II) can either be adsorbed (Demirbas, 2008) or precipitated as metal sulfide in anaerobic environments (Prasad et al., 2010). In contrast, their fate in selenium rich environment is poorly documented.

There is an increasing interest in the potential biotechnological applications of bacterial selenium oxyanion reduction as a green method for the production of metal selenide quantum dots (Ayano et al., 2013; Fellowes et al., 2013; Nancharaiah et al., 2015b). Among several metal selenides, mainly cadmium selenide (CdSe), zinc selenide (ZnSe) and lead selenide (PbSe) have attracted considerable attention due to their quantum confinement effects and size-dependent photoemission characteristics (Fellowes et al., 2013; Nancharaiah et al., 2015b). Hence, microbial reduction of selenium oxyanions in the presence of heavy metals (e.g. Cd, Zn, and Pb) is very important for the development of efficient bioremediation processes and for the microbial synthesis of metal selenide quantum dots (Nancharaiah et al., 2015b).

The use of sulfate reducing bacteria in metal bioremediation processes has been widely reported, for example, bioprecipitation as metal sulfide for cadmium (White et al., 1998), zinc and lead (Guo et al., 2010; Hien Hoa et al., 2007). However, to the best of our knowledge, there is no study on the effect of heavy metal co-contaminants on selenium oxyanion bioreduction or vice versa. Therefore, the objective of this work was to investigate microbial reduction of selenite in the presence of heavy metals. In this study, experiments were performed on selenite reduction by anaerobic granular sludge in the presence of different concentrations of three heavy metals, i.e. Cd(II), Zn(II) and Pb(II). Time course profiles of selenite removal along with the fate of bioreduced selenium and heavy metals were analyzed.

4.2. Materials and methods

4.2.1. Source of biomass

Anaerobic granular sludge was collected from a full-scale upflow anaerobic sludge blanket (UASB) reactor treating paper mill wastewater (Industriewater Eerbeek B.V., Eerbeek, The Netherlands) and was utilized as the inoculum for all experiments. The anaerobic granular sludge was characterized in detail by Roest et al. (2005). The sludge was stored at 4°C in an air tight jar under anaerobic conditions and used for selenite reduction experiments. All the experiments were performed in serum bottles under anaerobic conditions.

4.2.2.   Selenite reduction experiments

The mineral medium used in selenite reduction experiments contained (mg. L^{-1}): NH$_4$Cl (300), CaCl$_2$.2H$_2$O (15), KH$_2$PO$_4$ (250), Na$_2$HPO$_4$ (250), MgCl$_2$ (120), and KCl (250). Nitrilotriacetic acid (1 mg/mg heavy metal) was used as the chelating agent to prevent heavy metal precipitation. Sodium lactate (10 mM) was used as the carbon and electron source. Sodium selenite (Na$_2$SeO$_3$, 1 mM = 79 mg. L^{-1}) was used as the source of selenium. The pH of the medium was adjusted to 7.3 with 1 M NaOH. The medium was distributed into 100 mL volume glass serum bottles as 70 mL aliquots. The serum bottles were inoculated with 0.7 g wet weight (0.2 g dry weight) of anaerobic granular sludge. The bottles were purged with N$_2$ gas for ~5 min and incubated at 30°C on an orbital shaker set at 150 rpm for 9 d. All the experiments were performed either in duplicate or triplicate.

4.2.3.   Effect of heavy metals on selenite reduction

Selenite reduction experiments described above were performed in the presence of heavy metals. Stock solutions of heavy metals were prepared by dissolving 1 g. L^{-1} of CdCl$_2$, ZnCl$_2$, or PbCl$_2$ in ultrapure water. Heavy metals were added to serum bottles individually at different concentrations (10, 30, 50, 70, 90, 150, 300 and 400 mg. L^{-1}). To avoid precipitation, Pb concentrations were used up to a maximum of 150 mg. L^{-1}. Liquid samples were collected at regular time intervals for analyzing lactate, selenite, elemental selenium, total selenium and heavy metals. After 9 d of incubation, biomass was subjected to microwave-assisted acid digestion for measuring the total metal (Cd, Zn or Pb) and selenium concentration. Suitable controls were kept by incubating bottles containing only mineral medium and sodium lactate with selenite and heavy metals, but without the sludge biomass.

4.2.4.   Kinetics of heavy metal removal

The heavy metal (Cd, Zn and Pb) removal kinetics were analyzed by fitting the data using the pseudo-first order rate expression based on the solid capacity as given below:

$$\frac{dq_t}{dt} = k_1(q_e - q_t) \qquad\qquad (4.1)$$

Where q_e and q_t are the amount of metal biosorbed per unit weight of biosorbent (mg g^{-1} dry weight) at equilibrium and at any time t; respectively, and k_1 is the rate constant of pseudo-first order sorption (min^{-1}). After applying the initial and boundary conditions, for t = 0 and q_t = 0, the integrated form of the above equation becomes:

$$\log(q_e - q_t) = \frac{\log q_e - k_1}{2.303 \times t} \qquad (4.2)$$

In contrast to the pseudo-first order model, the pseudo-second order kinetic model predicts the behavior over the whole range of adsorption and is widely used by many researchers because it provides a more appropriate description than the first order equation (Volesky et al., 1995). It can be expressed in linear form as:

$$\frac{t}{q_t} = \frac{1}{k_2 q_e^2} + \frac{t}{q_e} \qquad (4.3)$$

Where q_t is the amount of the sorbate on the sorbent at time t (mg. g^{-1}), k_2 is the equilibrium rate constant of pseudo-second order sorption kinetics (g. mg^{-1} min^{-1}) and q_e is the equilibrium uptake (mg. g^{-1}) (Volesky et al., 1995). The pseudo-first and -second order constants were determined by plotting $\log(q_e - q_t)$ against t and t/q against t, respectively (Cordero et al., 2004).

4.2.5.   Analytical methods

The concentration of Cd, Zn and Pb was analyzed using an atomic absorption spectrophotometer (AAS) (PerkinElmer Model Analyst 200). Liquid samples were first filtered through a 0.45 µm cellulose acetate syringe filter (Sigma Aldrich, USA) and then the filtrate was analyzed for residual heavy metals after acidifying with concentrated nitric acid (pH<2) to prevent metal precipitation and adsorption onto surfaces.

For Se (IV) analysis, a modified spectrophotometric method was followed based on the method as described by Dao-bo et al. (2013). Liquid samples collected at different time points were centrifuged at 37,000 g for 15 min to remove the suspended cells and Se(0) particles. The supernatant (1 mL) was mixed with 0.5 mL of 4 M HCl, and then with 1 mL of 1 M ascorbic acid. After 10 min of incubation at room temperature, the absorbance was determined at 500 nm using an UV-Vis spectrophotometer (Hermle Z36 HK).

Se(0) was collected from the liquid phase by centrifuging at 37000 g for 15 min. After centrifugation, the supernatant was separated and the total selenium concentration in the supernatant was measured using a graphite furnace AAS (SOLAAR MQZe, unity lab services USA). The pellet was re-suspended in Milli-Q water and the Se concentration was determined using the same graphite furnace AAS. Aqueous selenide (HSe^-) concentration was calculated by subtracting concentrations of selenite in solution from the total Se (selenite + selenide) in solution by following Pearce et al. (2009).

4.3. Results

4.3.1. Effect of Cd, Zn and Pb on selenite reduction by anaerobic granular sludge

Selenite concentrations were monitored as a function of time for the anaerobic granular sludge in the presence of Pb, Zn and Cd separately at different concentrations and in the absence of heavy metals (Fig. 4.1A-C). At lower initial concentrations of Pb and Zn (10-70 mg. L^{-1}), selenite was completely reduced within 5 d of incubation. At higher initial concentrations of Pb and Zn (>90 mg. L^{-1}), selenite removal was not complete, but more than 92% was still removed (Fig. 4.1B, 1C). Overall, the presence of Pb and Zn in the medium did not exert a significant inhibition on the selenite reduction. In the presence of Cd, complete reduction of selenite was observed in 7 d at initial concentrations of 10-70 mg. L^{-1}. However, Cd showed a strong inhibition on selenite reduction at initial concentrations higher than 150 mg. L^{-1} and approximately 65-48% of 1 mM selenite was reduced (Fig. 4.1A).

Fig. 4.1(D-F) shows lactate consumption in the presence of different initial heavy metal concentrations. There was no significant effect of Pb and Zn on the lactate consumption at any of the Pb and Zn concentrations investigated. Also, Cd did not show any significant effect on lactate consumption at low initial concentrations (10-30 mg. L^{-1}). However, with an increase in initial Cd concentration (70-400 mg. L^{-1}), lactate consumption was inhibited: almost 40-80% lactate was not consumed and available in the medium, even at the end of the 9 d incubation period.

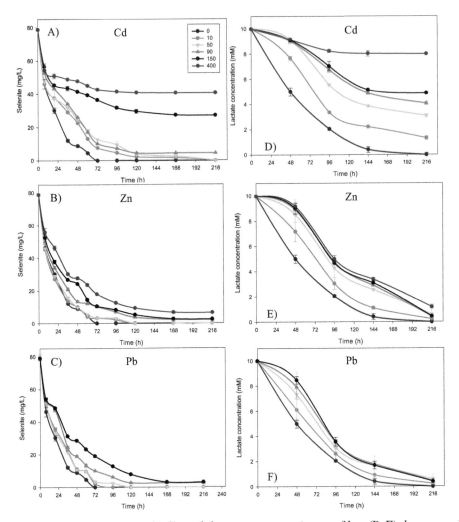

Fig. 4.1 Selenite reduction (A-C) and lactate consumption profiles (D-F) by anaerobic granular sludge in the presence of different concentrations of the heavy metals Pb, Zn, and Cd. To avoid precipitation, Pb was used only upto 150 mg. L^{-1}. Legend symbols are inside the panel (A)

The elemental selenium production was visually evident by the appearance of a red color during the incubation period. In the control incubation (without sludge biomass), there was no reduction of the selenite concentration and no change in color during the incubation period. Fig. 4.2A shows that production of red elemental Se was much higher in the presence of Cd as compared to Zn or Pb. It also shows that formation of elemental Se was increased with increased initial Cd concentration. However, the formation of elemental Se started decreasing again from 50 mg. L^{-1} Cd onwards. In contrast, production of elemental Se was much less at lower Pb and Zn concentrations and then started increasing at concentrations ≥90 mg. L^{-1}, but decreased again from 150 mg. L^{-1} onwards (Fig. 4.2A).

The concentration of aqueous selenide (HSe$^-$) was relatively higher in the presence of Cd as compared to Pb and Zn (Fig. 4.2B). Similar to elemental Se, the HSe$^-$ present in the liquid phase increased with the Cd concentration and reached a maximum at 50 mg. L^{-1} Cd and decreased thereafter. The production of HSe$^-$ was much less in the presence of Pb and Zn. At higher heavy metal concentrations (>150 mg. L^{-1}), the formation of HSe$^-$ was not observed.

The difference in selenium loading in the biomass under different conditions is shown in Fig. 4.3A. The selenium content of the biomass increased with an increase in the metal concentration up to 90 mg. L^{-1} and then gradually decreased with a further increase in metal concentration. For Pb, 90-80% of selenium was available in the biomass, while for Zn, the selenium concentration in the biomass decreased from 90 to 75% with increasing initial Zn concentration. The Cd concentration had the most significant effect on the selenium association with the biomass (Fig. 4.3A). A gradual increase in selenium concentration in the biomass from 50 to 80% was associated with an increase in initial Cd concentration from 10 - 90 mg. L^{-1}, and then the selenium concentrations in the biomass decreased at higher initial Cd concentrations. At 400 mg. L^{-1} Cd, the selenium concentration in the biomass was only 38% (Fig. 4.3A).

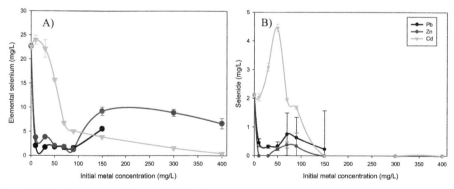

Fig. 4.2 Final concentrations of (A) elemental selenium and (B) selenide (HSe$^-$) during bioreduction of selenite by anaerobic granular sludge at the end of the 9d incubation period. Legend symbols are inside the panel (B)

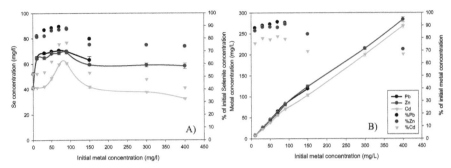

Fig. 4.3 Amount of (A) selenium and (B) heavy metals associated with anaerobic granular sludge in the presence of different concentrations of heavy metals Cd, Zn and Pb. Legend symbols are inside the panel (B)

4.3.2. Kinetics of heavy metal removal

The time course of Cd, Zn and Pb removal by anaerobic granular sludge is shown in Fig. 4.4. There was no significant effect of the initial metal concentration on the heavy metal removal efficiency at lower concentrations (10-90 mg. L^{-1}). After 9 d of incubation, the Pb and Zn concentrations were decreased to below 2 mg. L^{-1} in the liquid phase, resulting in more than 97% removal of metals. For Cd, more than 99% metal was removed from an initial concentration of 150 mg. L^{-1}. However, at a metal concentration of 150 mg. L^{-1}, the metal removal efficiency decreased to 86% for Pb, while removal efficiencies were 81% for Zn and 87% for Cd at 400 mg. L^{-1} metal concentration. It was also clear from Fig. 4.3B that 80-90% of the initial Pb was associated with the biomass. For Zn and Cd, although 80-90% of the metal was biosorbed at lower concentrations (10-90 mg. L^{-1}), the removal efficiency started to decline with a further increase in initial metal concentration.

In order to describe the biosorption process of Pb, Zn and Cd to the biomass, the kinetic data were described using the pseudo-first order and pseudo-second order rate equations. The correlation coefficients (R^2< 0.95) for the pseudo-first order kinetic constants were much lower than the pseudo-second order kinetic model (R^2 > 0.95), suggesting that the pseudo-second order kinetic model fitted the biosorption data better (Table 4.2). Table 4.2 shows the pseudo-second-order kinetic rate constant k$_2$ (pseudo-second order rate constant; mg. g^{-1} min^{-1}) and q$_e$ (sorption capacity; mg. g^{-1}) values, with the calculated q$_e$ being close to the experimental data, the pseudo-second order equation predicted equilibrium adsorption values more precisely than the first-order one. The biosorption capacity of the anaerobic granular sludge increased with increasing metal concentration. When the initial metal concentration increased from 10 to 400 mg. L^{-1}, the q_e increased from 1.01 to 34.84 mg g^{-1} for Cd and 1.03 to 32.89 for Zn. Similar results were observed in case of Pb: the biosorption capacity increased with an increase in initial metal concentration. However, for initial concentrations exceeding 150 mg. L^{-1} Pb, the metal cations began to precipitate in the medium prior addition to biomass. The rate parameter k_2 listed in Table 4.2 also indicated the biosorption of metals decreased at higher initial metal concentrations.

4.3.3. Selenium and heavy metal mass balances

The selenium mass balance was estimated from the obtained measurements at the end of the 9 d incubation. The mass balance was based on Se measurements and by assuming that the total initial selenium supplied as selenite (79 mg. L^{-1}) was converted partly to Se(0) and HSe$^-$ and partitioned between the liquid and biomass phases. Se in the liquid phase was distinguished in terms of residual selenite (Se(IV)), biogenic elemental Se (Se(0)) and selenide (HSe$^-$). The Se associated with the biomass was considered as total Se and no effort was made to distinguish between different selenium species, as this requires specialized speciation techniques as XANES and XAFS which were out of the scope of this study. The percentage ratio of total selenium partitioned between the liquid and biomass phase could be accounted to about 85-94% of the supplied selenium for all three heavy metals studied.

Similarly, the mass balance of Cd, Zn and Pb was determined. Heavy metal concentrations in the liquid phase were measured at regular intervals during the incubation period. However, the total metal concentration measured at the end of the incubation period includes the metal present in the medium and metal present in biomass. For Pb and Zn, the percent ratio of total final metal to the supplied initial metal was 88-94% while for Cd, it was varied from 80-87%.

4.4. Discussion

4.4.1. Effect of heavy metals on selenite reduction

This study shows that heavy metals, particularly Pb and Zn, did not significantly inhibit selenite reduction by anaerobic granular sludge at 30°C and pH 7.3 using lactate as electron donor. In contrast, Cd at higher concentrations (> 150 mg. L^{-1}) was inhibitory to the selenite bioreduction (Fig. 4.1A). The time-course of selenite removal revealed a pattern consisting of two distinct phases. At the beginning, the removal occurred rapidly indicating the biosorption of the selenite ion, followed by a second phase wherein removal was by the uptake and bioreduction of selenite by the biomass (Nancharaiah et al., 2015a). The inhibitory effect of Cd on selenite reduction confirms the more toxic nature of Cd to bacterial metabolism compared to Pb and Zn (Guo et al., 2010). At higher Cd concentrations (150 mg. L^{-1}), heavy metal removal was comparatively lower resulting in higher Cd(II) concentrations in the aqueous phase. The free metal ion concentration available in the aqueous phase eventually exerts toxicity to microorganisms and inhibits selenite bioreduction (Bartacek et al., 2008).

However, the toxicity of the heavy metal depends on several other factors, including metal type and speciation, type of microorganisms, species composition of microbial community and their defense mechanisms (Ayano et al., 2013; Bartacek et al., 2008; Kieu et al., 2011). For example, Cd, Zn and Pb at concentrations of 20, 25, and 75 mg. L^{-1}, respectively are toxic to sulfate reducing bacteria (SRB) in batch experiments (Hao et al., 1994). Another report on heavy metal bioremediation by a mixture of SRB showed that Zn had significantly less toxic effects on SRB with inhibition caused only at 400 mg. L^{-1} Zn (Zhou et al., 2013). This suggests that SRB are more sensitive to heavy metal toxicity than the selenite reducing population present in the anaerobic granular sludge investigated in this study.

There are no studies on the effect of heavy metals on anaerobic selenate or selenite reducing bacteria. Recently, selenite reduction and formation of CdSe by *Pseudomonas* sp. in the presence of 183 mg. L^{-1} Cd was reported under aerobic growth conditions (Ayano et al., 2013). *Pseudomonas* sp. isolated through an enrichment process was able to grow and carry out selenite reduction even at 3660 mg. L^{-1} Cd (Ayano et al., 2014). This is to the best of our knowledge the only study which investigated Cd resistance and selenite reduction in the presence of Cd using *Pseudomonas* sp. in aerobic conditions. No studies have been done so far on the effect of heavy metals on selenite reduction by pure or mixed cultures under anaerobic conditions.

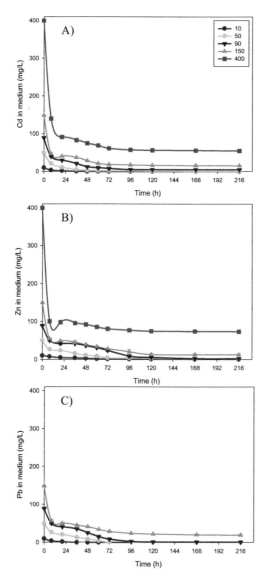

Fig. 4.4 Residual heavy metal concentration in the liquid media over time. A) Cd; B) Zn; and C) Pb. Legend symbols are inside the panel (A)

Table 4.2 Second-order kinetics constant for the biosorption of Pb, Zn and Cd on anaerobic granular sludge.

Initial metal concentration (mg. L^{-1})	q_e			k_2			R^2		
	Cd	Zn	Pb	Cd	Zn	Pb	Cd	Zn	Pb
10	1.017	1.03	1.00	0.41	0.07	0.47	0.99	0.99	0.99
30	3.04	3.16	3.18	0.12	0.02	0.08	0.99	0.99	0.98
50	5.04	5.04	5.21	0.06	0.02	0.02	0.99	0.99	0.99
70	6.73	7.29	7.26	0.03	0.01	0.02	0.99	0.97	0.99
90	8.71	9.50	9.48	0.02	0.00	0.01	0.99	0.97	0.98
150	14.29	14.16	13.33	0.02	0.01	0.01	0.99	0.99	0.99
300	26.31	26.04	-	0.02	0.01	-	0.99	0.99	-
400	34.84	32.89	-	0.01	0.02	-	0.99	0.99	-

q_e - Adsorbed metal ion quantity per gram of biomass at equilibrium (mg g^{-1})

k_2 - Second-order adsorption rate constant (g mg^{-1}min^{-1})

R^2 - Correlation coefficient

4.4.2.   Fate of selenium

Formation of Se species such as Se(0) and HSe$^-$ were calculated during selenate bioreduction in the presence of heavy metals. The amount of Se (Se(0) + HSe$^-$) was higher in the liquid phase when the medium contained Cd (Fig. 4.2). This is possible because of the formation of CdSe colloids which remain in the aqueous phase due to their size and surface properties. The same was not observed in the presence of Zn and Pb, because the formation of CdSe is thermodynamically more favored than ZnSe and PbSe formation (Pearce et al., 2008). In addition, the Cd-Se bond strength in CdSe is much higher than the Zn-Se bond strength in ZnSe (Sung et al., 2006). In an analogous sulfide system, it was suggested that Zn is more likely than Cd to form oxide/hydroxide phases, because of the relatively small difference in the solubility products of ZnS and ZnO/Zn(OH)$_2$ (O'Brien et al., 1998). Pb has a similar chemical behavior like Zn. Under the experimental conditions, formation of Pb carbonate or oxide precipitates is more likely rather than sulfide or selenide phases (Hesterberg et al., 1997; O'Day et al., 2000). However, in the presence of 90 mg. L^{-1} Pb and Zn onwards, an increase in elemental Se and HSe$^-$ was observed (Fig. 4.2). It is likely that higher concentrations of Zn and Pb interfered with the selenite removal mechanisms, particularly with biosorption. Thus, more selenite present in the liquid phase was converted to relatively higher amounts of Se(0) and Se(-II) in the aqueous phase by bioreduction by the microorganisms present in the anaerobic granular sludge.

Simulation by Visual MINTEQ software confirmed the formation of HSe⁻, which reacts with Cd, Zn or Pb available in the medium to form CdSe, ZnSe or PbSe, respectively. Charge differences of all the simulation models were less than 5% which supports the applicability for the experimental system. According to the simulation, all the HSe⁻ is used to form metal selenides when the aqueous heavy metal concentration equals or exceeds the available HSe⁻. A similar behavior for consumption of biogenic sulfide for heavy metal precipitation as metal sulfides was reported by Villa-Gomez et al. (2012).

From Fig. 4.3 it is clear that more than 90% of the added Pb and Zn along with selenium were present in the anaerobic granular sludge, which could be in the form of PbSe or ZnSe, respectively. Another reason could be that a large amount of selenite was reduced rapidly to elemental selenium and the heavy metals adsorbed on these Se(0) nanoparticles lead to the precipitation of elemental selenium (Jain et al., 2015).

4.4.3.   Fate of heavy metals

The time-course measurements (Fig. 4.4) demonstrated that removal of heavy metals by anaerobic granular sludge occurred rapidly at the beginning through biosorption, which is a spontaneous process and thereby often occurs very fast (Volesky, 2001; Yuan et al., 2009). It also reveals that the bioremoval of heavy metals had a second removal phase probably via intracellular accumulation (Yuan et al., 2009) or as metal selenide precipitate (Fellowes et al., 2013) or sorption onto Se(0) (Jain et al., 2015). Indeed, the heavy metals are not inhibiting the microbial metabolism, as Se bioreduction continued during the incubation in the presence of even high initial heavy metals concentrations (Fig. 4.1A-C).

Heavy metal removal data fitted well using the pseudo-second order equation with R^2 values of > 0.99 (Table 4.2). The results predict the behavior over the entire study range, with a chemisorption mechanism involving valency forces through sharing or exchange of electrons between sorbent and sorbate being the rate controlling step (Choi et al., 2009). The experimental results showed an increase in metal adsorption capacity (Table 4.2) while the removal efficiency decreased with an increase in initial metal ion concentration (Fig. 4.4). The initial increase in metal adsorption capacity resulted from the increase in driving force, i.e., concentration gradient due to adsorption. The decrease in the heavy metal removal efficiency at elevated concentrations might be due to the saturation of binding sites for adsorption (King et al., 2008; Yang et al., 2010). The rate parameter K_2 (Table 4.2) derived from fitting experimental data indicated that the biosorption at higher metal concentrations is slower due to the competition for binding sites.

4.4.4. Proposed selenium removal mechanism in the presence of heavy metals

Based on the results presented above, a mechanism for the removal of selenite and heavy metals through different processes is proposed (Fig. 4.5). Granular sludge became red due to the formation of Se(0) in the granular sludge and precipitation of Se(0) formed in the aqueous phase by bioreduction of selenite in the presence of heavy metals, using lactate as electron donor. The removal of heavy metals is mainly driven by biosorption onto the granular sludge (Volesky, 2001; Yuan et al., 2009). Adsorption of heavy metals onto Se(0) nanoparticles can partly contribute to the removal of the heavy metals from aqueous phase as well. Subsequent reduction of Se(0) will form HSe⁻ which precipitates with heavy metals to form metal selenides. Depending on the size and surface properties, the formation of metal selenides both in the granular sludge and in the aqueous phase is expected.

To further evaluate the chemical environment of heavy metals, especially whether the metals are sorbed onto the Se(0) nanoparticles or bound as metal selenides (e.g. CdSe), further speciation studies are required using X-ray absorption spectroscopic techniques e.g. XANES and EXAFS. Nevertheless, the results presented in this paper show for the first time the effect of heavy metal co-contaminants on the microbial selenite reduction and the fate of bioreduced selenium in anaerobic granular sludge.

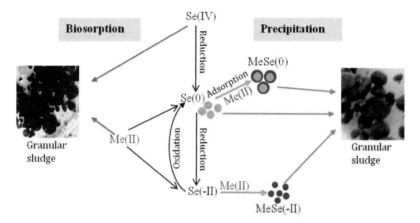

Fig. 4.5 Proposed mechanism for removal of selenium and heavy metal by anaerobic granular sludge. Granular sludge became red due to the precipitation of mainly red elemental Se, formed during the bioreduction of selenite in the presence of heavy metals like Cd, Zn or Pb. Precipitation of metal sorbed onto Se(0) and/or metal selenide is also responsible for selenium as well as metal removal. Me(II): Heavy metal (Cd, Zn or Pb), MeSe(0): Metal sorbed onto Se(0), MeSe(-II): Metal selenide

4.4.5. Practical implications

Heavy metals are common co-contaminants with selenium oxyanions in different selenium containing waste streams (Table 4.1). The present study demonstrated that bioreduction of selenite by anaerobic granular sludge is possible, even in the presence of high (up to 400 mg. L^{-1}) concentrations of heavy metals. The results obtained here are useful for the development of a selenium oxyanion-reducing bioreactor to simultaneously remove selenium oxyanions and heavy metals from wastewaters. The majority of studies on selenium wastewaters noted the presence of colloidal Se(0) in the effluents of bioreactors, necessitating post-treatment prior to discharge. It is to be noted that this information is based on microbial reduction of selenium oxyanions performed in the absence of heavy metal co-contaminants.

In this study, most of the selenium and heavy metals was found to be associated with the biomass (~90%), which will increase the Se removal efficiency of the UASB bioreactor and will thus make the post-treatment redundant. It is desirable to investigate the treatment of selenium oxyanions in bioreactors in the presence of heavy metal co-contaminants in order to ascertain the fate of Se under long term operating conditions. Furthermore, the anaerobic sludge was able to reduce selenite up to selenide in the presence of heavy metals, which opens perspectives to develop a novel microbial synthesis process to produce metal selenide nanoparticles by combining bioremediation approaches to convert environmentally problematic waste streams to 'high-end' saleable products like metal chalcogenide quantum dots (Pearce et al., 2008).

4.5. Conclusion

This work showed the effect of Cd, Zn and Pb on the microbial reduction of selenite. Both elemental selenium and selenide were the end products. Cd showed an inhibitory effect on selenite bioreduction at a concentration higher than 150 mg. L^{-1} Cd. In contrast, Pb and Zn did not exert a significant effect on selenite bioreduction. Production of aqueous fractions of Se(0) and HSe^- were much higher in the presence of Cd as compared to Pb and Zn. The majority of the bioreduced selenium and heavy metals were associated with the anaerobic granular sludge.

Acknowledgements

This research was supported through the Erasmus Mundus Joint Doctorate Environmental Technologies for Contaminated Solids, Soils, and Sediments (ETeCoS³) (FPA n⁰ 2010-0009) and BioMatch project No. 103922, funded by the European Commission Marie Curie International Incoming Fellowship (MC-IIF).

References

Ayano, H., Kuroda, M., Soda, S., Ike, M. 2014. Effects of culture conditions of *Pseudomonas aeruginosa* strain RB on the synthesis of CdSe nanoparticles. *J Biosci Bioeng.* 119(4), 440-445.

Ayano, H., Miyake, M., Terasawa, K., Kuroda, M., Soda, S., Sakaguchi, T., Ike, M. 2013. Isolation of a selenitereducing and cadmium-resistant bacterium *Pseudomonas sp.* strain RB for microbial synthesis of CdSe nanoparticles. *J Biosci Bioeng.* 117(5), 576-581.

Bartacek, J., Fermoso, F.G., Baldó-Urrutia, A.M., van Hullebusch, E.D., Lens P.N.L. 2008. Cobalt toxicity in anaerobic granular sludge: influence of chemical speciation. J. Ind. *Microbiol Biotechnol.* 35(11), 1465-1474.

Beyenal, H., Lewandowski, Z. 2004. Dynamics of lead immobilization in sulfate reducing biofilms. *Water Res.* 38, 2726-2736.

Choi, J., Lee, Y.J., Yang, J-S. 2009. Biosorption of heavy metals and uranium by starfish and *Pseudomonas putida. J Hazard Mater.* 161(1), 157-162.

Cordero. B., Lodeiro, P., Herrero, R., Sastre de Vicente, E.M. 2004. Biosorption of cadmium by fucus spiralis. *Environ Chem.* 1(3), 180-187.

Dao-Bo. L., Yuan-Yuan, C., Chao, W., Wen-Wei, L., Na, L., Zong-Chuang, Y., Zhong-Hua, T., Han-Qing, Y. 2013. Selenite reduction by *Shewanella oneidensis* MR-1 is mediated by fumarate reductase in periplasm. *Sci Rep.* 4, 1-7.

Demirbas, A. 2008. Heavy metal adsorption onto agro-based waste materials: a review. *J Hazard Mater.* 157(2-3), 220-229.

España, J.S., Pamo, E.L., Pastor, E.S., Andrés, J.R., Rubí, J.A.M. 2006. The Impact of acid mine drainage on the water quality of the odiel river (Huelva, Spain): evolution of precipitate mineralogy and aqueous geochemistry along the Concepción-Tintillo segment. *Water Air. Soil Pollut.* 173, 121 - 149.

Fellowes, J.W., Pattrick, R.A.D., Lloyd, J.R., Charnock, J.M., Coker, V.S., Mosselmans, W., Weng, T.C., Pearce, C.I. 2013. Ex situ formation of metal selenide quantum dots using bacterially derived selenide precursors. *Nanotechnology* 24(14), 145603 -145612.

Guo, H, Luo, S., Chen, L., Xiao, X., Xi, Q., Wei, W., Zeng, G., He, Y. 2010. Bioremediation of heavy metals by growing hyperaccumulaor endophytic bacterium *Bacillus sp.* L14. *Bioresour Technol.* 101, 8599-8605.

Hao, O.J., Huang, L., Chen, J.M., Bugass, R.L., 1994. Effects of metal additions on sulfate reduction activity in wastewater. *Toxicol Environ. Chem.* 46(4), 197-212.

Hesterberg, D., Sayers, D.E., Zhou, W.Q., Plummer, G.M., Robarge, W.P. 1997. X-ray absorption spectroscopy of lead and zinc speciation in a contaminated groundwater aquifer. *Environ Sci Technol.* 31, 2840-2846.

Hien Hoa, T.T., Liamleam, W., Annachhatre, A.P. 2007. Lead removal through biological sulfate reduction process. *Bioresour Technol.* 98(13), 2538-2548.

Jain, R., Jordan, N., Schild, D., van Hullebusch, D.E., Weiss, S., Franzen, C., Farges, F., Hübner, R., Lens, P.N.L. 2015. Adsorption of zinc by biogenic elemental selenium nanoparticles. *Chem Eng J.* 260, 855-863.

Kieu, T.Q.H., Muller, E., Horn, H. 2011. Heavy metal removal in anaerobic semi-continuous stirred tank reactors by a consortium of sulfate-reducing bacteria. *Water Res.* 45, 3863-3870.

King, P., Rakesh, N., Lahari, B.S., Kumar, P.Y., Prasad, V.S.R.K. 2008. Biosorption of zinc onto *Syzygium cumini* L.: Equilibrium and kinetic studies. *Chem Eng J.* 144(2), 181-187.

Lenz, M., Lens, P.N.L. 2009. The essential toxin: The changing perception of selenium in environmental sciences. *Sci Total Environ.* 407(12), 3620-3633.

Lenz, M., van Hullebusch, D.E., Hommes, G., Corvini, P.F.X., Lens, P.N.L. 2008. Selenate removal in methanogenic and sulfate-reducing upflow anaerobic sludge bed reactors. *Water Res.* 42(8-9), 2184-2194.

Matlock, M.M., Howerton, B.S., Atwood, D.A. 2002. Chemical precipitation of heavy metals from acid mine drainage. *Water Res.* 39, 4757 - 4764.

Meawad, A.S., Bojinova, D.Y., Pelovski, Y.G. 2010. An overview of metals recovery from thermal power plant solid wastes. *Waste Manage.* 30(12), 2548 - 2559.

Moreau, J.W., Fournelle, J.H., Banfield, J.F. 2013. Quantifying heavy metals sequestration by sulfate-reducing bacteria in an acid mine drainage-contaminated natural wetland. *Front Microbiol.* 4, 1 - 10.

Nancharaiah, Y.V., Lens, P.N.L. 2015a. Ecology and biotechnology of selenium-respiring bacteria. *Microbiol Mol Biol Rev.* 79, 61-80.

Nancharaiah, Y.V., Lens, P.N.L. 2015b. Selenium biomineralization for biotechnological applications. *Trends Biotechnol.* 33, 323-330.

Nancharaiah, Y.V., Venkata Mohan, S., Lens, P.N.L. 2015c. Metal removal and recovery in microbial fuel cells: a review. *Bioresour Technol.* 195, 102-114.

O'Brien, P., McAleese, J. 1998. Developing an understanding of the processes controlling the chemical bath deposition of ZnS and CdS. *J Mater Chem.* 8, 2309-2314.

O'Day, P.A., Carroll, S.A., Randall, S., Martinelli, R.E., Anderson, S.L., Jelinski, J., Knezovich, J.P. 2000. Metal speciation and bioavailability in contaminated estuary sediments, Alameda Naval Air Station, California. *Environ Sci Technol.* 34, 3665-3673.

Pearce, C.I., Coker, V.S., Charnock, J.M., Pattrick, R.A.D., Mosselmans, J., Law, N., Beveridge, T.J., Lloyd, J.R. 2008. Microbial manufacture of chalcogenide-based nanoparticles via the reduction of selenite using *Veillonella atypica*: an *in situ* EXAFS study. *Nanotechnology* 19, 155603-155615.

Pearce, C.I., Pattrick, R.A.D., Law, N., Charnock, J.M., Coker, V.S., Fellowes, W., Oremland, R.S., Lloyd, J.R. 2009. Investigating different mechanisms for biogenic selenite transformations: *Geobacter sulfurreducens*, *Shewanella oneidensis* and *Veillonella atypica*. *Environ Technol.* 30(12), 1313-1326.

Prasad, k., Jha, A.K. 2010. Biosynthesis of CdS nanoparticles: An improved green and rapid procedure. *J Colloid Interface Sci*. 342, 68-72.

Roest, K., Heilig, H.G.H.J., Smidt, H., Vos de, W.M., Stams, A.J.M., Akkermans, A.D.L. 2005. Community analysis of a full-scale anaerobic bioreactor treating paper mill wastewater. *Syst Appl Microbiol*. 28, 175 -185.

Sung,Y.M., Lee, Y.J., Park, K.S. 2006. Kinetic analysis for formation of $Cd_{1-x}Zn_xSe$ solid-solution nanocrystals. *J Am Chem Soc*. 128(28), 9002-9013.

Villa-Gomez, D.K., Papirio, S., van Hullebusch, E.D., Farges, F., Nikitenko, S., Kramer, H., Lens, P.N.L. 2012. Influence of sulfide concentration and macronutrients on the characteristics of metal precipitates relevant to metal recovery in bioreactors. *Bioresour Technol*. 110, 26-34.

Volesky, B. 2001. Detoxification of metal-bearing effluents: biosorption for the next century. *Hydrometallurgy* 59, 203-216.

Volesky, B., Holan, Z.R. 1995. Biosorption of heavy metals. Biotechnol. Prog. 11(3), 235-250.

White, C., Gadd, G.M. 1998. Accumulation and effects of cadmium on sulfate-reducing bacterial biofilms. *Microbiology* 144, 1407-1415.

Yang, C., Wang, J., Lei, M., Xie, G., Zeng, G., Luo, S. 2010. Biosorption of zinc(II) from aqueous solution by dried activated sludge. *J Environ Sci*. 22(5), 675-680.

Yuan, H.P., Zhang, J.H., Lu, Z.M., Min H., Wu C. 2009. Studies on biosorption equilibrium and kinetics of Cd^{2+} by *Streptomyces sp*. K33 and HL-12. *J Hazard Mater*. 164, 423-431.

Zhou, Q., Chen, Y., Yang, M., Li, W., Deng, L. 2013. Enhanced bioremediation of heavy metal from effluent by sulfate-reducing bacteria with copper–iron bimetallic particles support. *Bioresour Technol*. 136, 413-417.

CHAPTER 5

Biosynthesis of CdSe nanoparticles by anaerobic granular sludge

This chapter has been published in modified form:

Mal, J., Nancharaiah, Y.V., van Hullebusch, Bera, S., Maheshwari, N., Lens, P.N.L. 2017. Biosynthesis of CdSe nanoparticles by anaerobic granular sludge. *Environ Sci Nano*. 4, 824-833

Chapter 5

Abstract

This study investigatedthe feasibility of combining bioremediation of selenium (Se) containing wastewater and biorecovery of Se as cadmium selenide nanoparticles (CdSe NPs). The microbial community ofanaerobic granular sludge was enriched for 300 days in the presence of Cd(II) and selenite (Se(IV)). Complete Se(IV) (79mg. L^{-1}) reduction in the presence of Cd (30 mg. L^{-1}) was observed after the 16th transfer (8 months) with the formation of both elemental selenium (Se(0)) and dissolvedselenide (Se(-II)). Cd was either associated with Se(0) orremained in the aqueous phase as free Cd(II) ion and/or as CdSe. The absorption and the fluorescence spectra of the aqueous phase showedformation of CdSe NPs. UV-vis and X-ray photoelectron spectroscopy (XPS) confirmed that the CdSe NPs were capped by extracellular polymeric substances (EPS) originating from the anaerobic granular sludge. Raman spectroscopy and XPS analysis further confirmed the presence of CdSe NPs in the aqueous phase, while Cd present inthe Se(0) pellet after centrifugation was mainly precipitated as a Se(0)-Cd complex. ACdSe/CdS core/shell structure was found in the sludge, suggesting thatCd(II) ions on the surface of the CdSe core interact with the sulfhydryl (-SH) groups present in the EPS of the UASB granules.

Keywords: Cadmium selenide; Selenite reduction; Elemental selenium (Se(0)); Cd-tolerant selenite-reducing consortium, Biogenic nanoparticles

5.1. Introduction

Cadmium selenide quantum dots (CdSe QDs) have attracted considerable interest for their potential applications in solar cells and optoelectronic sensors (Mal et al., 2016a). CdSe QDs have a size range of 1–20 nm and are widely used as inorganic fluorophores in the field of biology and medicine for imaging and sensing, including for fluorescent biolabelling and cancer detection (Mal et al., 2016a). Various physical and chemical methods are available for synthesizing CdSe QDs. However, the development of novel green and environmental friendly methods which use renewable materials instead of toxic and hazardous chemicals is desired for synthesizing CdSe QDs (Ayano et al., 2013; Fellowes et al., 2013; Mal et al., 2016a).

Microbial reduction of selenium-oxyanions (i.e. selenate and selenite) into elemental selenium (Se(0)) and selenide (Se(-II)) plays an important role in the biogeochemical selenium cycling and detoxification of soluble and toxic Se-oxyanions in the natural environment (Dessì et al., 2016; Nancharaiah & Lens, 2015). In environmental settings, selenium is often found as metal selenide minerals in rocks and sediments under highly reducing conditions, but not as elemental selenium (Boyd, 2011; Nancharaiah & Lens, 2015). Selenide can be incorporated into natural sulfide minerals (i.e. pyrite (FeS_2), chalcopyrite ($CuFeS_2$), and sphalerite (ZnS)) due to its chemical similarity to sulphur (Nancharaiah & Lens, 2015). To the best of our knowledge, the role of microorganisms in the formation of metal selenides in environmental settings is unknown. Heavy metals like Cd andZn are common co-contaminants found along with selenium oxyanions in different selenium containing waste streams (Mal et al., 2016b). The presence of these heavy metals, either in selenium-rich wastewaters or in environmental settings, may thus influence the microbial reduction processes as well as the speciation of bioreduced selenium (Mal et al., 2016b). Recently, we have reported that the formation of biogenic Se(0) or aqueous selenide (Se(-II)) was influencedby the presence of heavy metals(cadmium, zinc and lead) and their concentration (Mal et al., 2016b). Biogenic Se (Se(0) and Se(-II)) concentrations were found to be higher in the liquid phase when the microbial reduction of selenite was performed in the presence of Cd. One possibility for this observation was the formation of CdSe colloids which can remain in suspension due to their smaller size and surface characteristics. Based on these previous results (Mal et al., 2016b), the present paper focuses on the potential formation of CdSe nanostructures during microbial reduction of selenium oxyanions in the presence of Cd.

Till to-date, there are only a few reports on the formation of CdSe nanoparticles (NPs) during microbial reduction of selenium oxyanions (Ayano et al., 2014; Fellowes et al., 2013; Yan et al., 2014). Particularly, studies on the formation of CdSe NPs coupled to bioremediation of Se-containing wastewaters are limited, while the majority of the previous studies on fungal based CdSe synthesis employed (externally supplied) selenium tetrachloride ($SeCl_4$) as the Se source (Kumar et al., 2007; Suresh, 2014). Since selenium in the environment and wastewaters principally exists in the form of oxyanions (selenite or selenate), the use of these Se-oxyanions as precursors for green synthesis of CdSe QDs is more appealing as wastewater treatment and bioremediation can then be coupled to Se recovery and the production of value added materials, i.e. CdSe QDs (Mal et al., 2016a).

Being toxic to microorganisms, Cd can impact biological treatment of selenium contaminated wastewater (Ayano et al., 2013; Mal et al., 2016b). It is, therefore, important to engineer the microbial community capable to be both cadmium tolerant and able to reduce Se(IV) to Se(-II) at the same time. The overall objective of this work was to utilize a selenium reducing microbial community for removing Se(IV) from wastewater and to recover Se in the form of CdSe NPs. In the present study, we describe the enrichment of a microbial community present in anaerobic granular sludge capable of reducing selenite to selenide in the presence of Cd for the synthesis of CdSe NPs. Special attention was given to understand the cadmium-selenium interaction and selenium fractionation using Raman spectroscopy and X-ray photoelectron spectroscopy (XPS).

5.2. Materials and Methods

5.2.1. Source of biomass

Anaerobic granular sludge was collected from a full scale upflow anaerobic sludge blanket (UASB) reactor treating paper mill wastewater (Industriewater Eerbeek B.V., Eerbeek, The Netherlands). A detailed description of the anaerobic granular sludge was given by Roest et al. (2005). The collected sludge was storedat 4°C in an air tight jar under anaerobic conditions and later used as inoculum for studying the selenite reduction in the presence of cadmium and synthesis of CdSe NPs (Mal et al., 2016b).

5.2.2.   Enrichment of granular sludge for selenite reduction in the presence of Cd(II)

The mineral medium used in this study contained (mg. L^{-1}): NH$_4$Cl (300), CaCl$_2$.2H$_2$O (15), KH$_2$PO$_4$ (250), Na$_2$HPO$_4$ (250), MgCl$_2$ (120), and KCl (250). Sodium lactate (5mM) was used as the carbon source and sodium selenite (1 mM = 79 mg. L^{-1}) was used as selenite source. The pH of the medium was adjusted to 7.3 with 1 M NaOH. The medium was distributed into 500 mL volume glass serum bottles as 400 mL aliquots. The serum bottles were inoculated with 3.2 g wet weight (1.0 g dry weight) of anaerobic granular sludge. After 2 weeks of incubation, the granular sludge was transferred to the new medium. This operation was repeated for a total of 20 transfers over 300 days to enrich the microbial community capable of reducing selenite to selenide in the presence of Cd and synthesizing CdSe.

The stock solution of Cd(II) was prepared by dissolving 1 g. L^{-1} of CdCl$_2$ in ultrapure water. The initial Cd concentration in serum bottles was gradually increased from 10 to 50 mg. L^{-1} and then later decreased to 30 mg. L^{-1}. Nitrilotriacetic acid (1mg. mg^{-1} Cd) was used as the chelating agent to prevent the precipitation of Cd(II) in the medium. The bottles were purged with N$_2$ gas for ~5 min and incubated at 30°C on an orbital shaker set at 180 rpm. After every incubation period, liquid samples were collected for analysing lactate, Se(IV), total selenium, Se(0), Se(-II) and Cd.

5.2.3.   Particle size distribution of selenium nanoparticles in aqueous phase

Liquid samples collected at the end of every incubation period were subjected first to centrifugation at 37000 g for 20 min to separate the Se(0) particles from the aqueous selenide (Se(-II)) and/or colloidal CdSe NPs (Mal et al., 2016b). The granular sludge that remained in the serum bottles was referred to as phase I (Fig. 5.1). The pellet (phase II) collected from the centrifugation step was re-dispersed in Milli-Q water and the supernatant (phase III) was used for size distribution measurements by using dynamic light scattering (DLS) performed on a Zetasizer Nano-ZS instrument (Malvern Instruments) (Pawar et al., 2013). Transmission electron microscopy (TEM) was performed on a JEOL JEM 1200 EX TEM machine operated at an accelerating voltage of 200 kV.

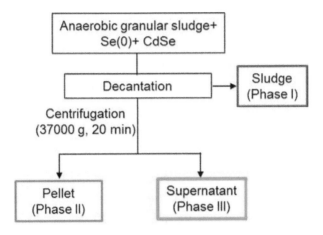

Fig. 5.1 Separation of three phases of Selenium for detailed selenium speciation

5.2.4.   Absorbance and fluorescence measurements

Aliquots from both the pellet (phase II) and supernatant (phase III) were subjected to UV-Vis spectrophotometric measurements by recording the absorbance spectra (280 - 600 nm) at a resolution of 1 nm performed on a Shimadzu dual-beam spectrophotometer (model UV-1601 PC). The supernatant sample was filtered through a 0.05 μm cellulose acetate syringe filter (Sigma Aldrich, USA) to remove any remaining Se(0) from the sample before taking UV-Vis and fluorescence spectra. The fluorescence measurements carried out using Fluoromax-3 (HORIBA Jobin Yvon Inc). The samples were excited at 365 nm and the emission spectra were recorded from 400 - 650 nm (Fellowes et al., 2013).

5.2.5.   Characterization of biosynthesized CdSe nanoparticles

Sludge samples (phase I) collected after the enrichment, the pellet (phase II) and the supernatant (phase III) fractionated from the aqueous phase (Fig. 5.1) were dried at 30°C and powdered for XPS and Raman spectroscopy analysis.Powder samples were pressed on indium foil for XPS analysis (Model VG ESCA LAB MK 200 ×, UK). An Al k$_\alpha$ X-ray source (energy 1486.6 eV) was used for the experiment. The high resolution spectra were collected at 20 eV pass energy for Se and 50 eV pass energy for Cd. Micro-Raman spectroscopy (Renishaw inVia Raman microscope, United Kingdom) was performed on all three samples to unambiguously distinguish the presence of Se(0) and CdSe. With this method, a region of interest was identified in the optical image and the microscope laser beam was focused. The excitation wavelength was 532 nm, produced by a Nd-YAG laser and the Raman spectrum was recorded typically using a 2 μm diameter beam.

5.2.6. Analytical methods

Total selenium concentration in the liquid phase collected after the end of every incubation period was measured using a graphite furnace atomic absorption spectrophotometer (AAS) (SOLAAR MQZ, unity lab services USA) after acidifying with concentrated nitric acid as described by Mal et al. (2016b). Then liquid samples were centrifuged (Hermle Z36 HK) at 37000 g for 20 min. Se(0) collected in the pellet was re-suspended in Milli-Q water. The total Se concentration in the pellet (phase II) and supernatant (phase III) was determined using the graphite furnace AAS.

For Se(IV) analysis, a modified spectrophotometric method was followed based on the method as described by Mal et al. (2016b). Briefly, supernatant (1 mL) was mixed with 0.5 mL of 4 M HCl, and then with 1 mL of 1 M ascorbic acid.After 10 min of incubation at room temperature, the absorbance was determined at 500 nm using an UV-Vis spectrophotometer. The aqueous selenide concentration was calculated by subtracting the selenite concentration in solution from the total Se (selenite + selenide) concentration following Pearce et al. (2008).

The Cd concentration in the pellet (phase II) and supernatant (phase III) was analysed to measure, respectively, precipitated Cd and residual Cd ions (or as CdSe) using AAS (Perkin Elmer Model Analyst 200). The samples were acidified with concentrated nitric acid (pH<2) to prevent metal precipitation and adsorption onto surfaces before performing AAS.

5.3. Results

5.3.1. Enrichment of granular sludge for aqueous selenide formation for CdSe synthesis

The selenite reduction and fractionation of selenium in the presence of cadmium after each incubation period during the enrichment of the anaerobic granular sludge is shown in Fig. 5.2. Initially, Cd at 10 and 20 mg. L^{-1} did not exert significant inhibition on selenite reduction:

complete removal of selenite via reduction was noticed. However, when the initial Cd concentration was increased to 50 mg. L^{-1}, selenite reduction was negatively affected and was not complete (Fig. 5.2A). Adding to that, the selenite reduction efficiency kept decreasing in the successive batches in the presence of 50 mg. L^{-1} of Cd. About 15% of 1 mM selenite was not reduced and remained in the medium after the 8th transfer (Fig. 5.2A). In order to improve the selenite reduction efficiency, the Cd concentration was decreased to 20 mg. L^{-1}. Fig. 5.2A shows that this operational change has gradually improved the selenite reduction efficiency in subsequent batches. As a result, only about 3% of 1 mM selenite remained unreduced after the 12th successive transfer. Then, the Cd concentration was increased to 30 mg. L^{-1} which did not impact the selenite reduction efficiency. After the 16th transfer, complete reduction of selenite was observed in the presence of 30 mg. L^{-1} Cd, which continued until the last day of the enrichment process (Fig. 5.2A).

Not only selenite reduction, also the formation of Se(0) and Se(-II) in the aqueous phase was altered with the change in Cd concentration (Fig. 5.2A). Concentrations of both Se(0) and Se(-II) increased with every transfer of medium at the beginning in the presence of low Cd concentrations (10 and 20 mg. L^{-1}). Se(0) and Se(-II)concentrations were found to be about 19.3 and 1.8 mg. L^{-1}, respectively, at the end of the 4th transfer (Fig. 5.2A). However, with the increase in Cd concentration to 50 mg. L^{-1}, concentrations of Se(0) and Se(-II) gradually decreased after every incubation period. After the 8th transfer, the Se(0) and Se(-II) concentration decreased to about 11 and 0.3 mg. L^{-1}, respectively (Fig. 5.2A). The Se(0) and Se(-II) concentration, however, increased again when the Cd concentration was decreased. After the 16th transfer, formation of Se(0) and Se(-II) reached to a maximum level and became stable. At the end of the enrichment period, the Se(0) and Se(-II) concentration in the aqueous phase was about 25.9 and 8 mg. L^{-1}, respectively (Fig. 5.2A).

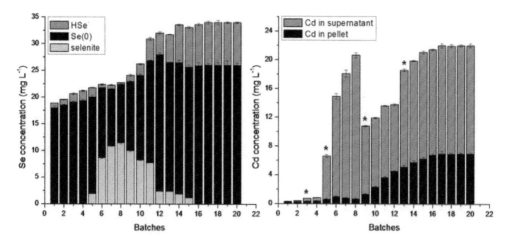

*Fig. 5.2 Time course of **A)** selenite reduction and selenium speciation in the presence of Cd and **B)** cadmium removal during the enrichment of anaerobic granular sludge. Initial concentration of selenite was 1 mM (79 mg. L⁻¹). (*) indicates the change in initial Cd concentration as 10, 20, 50, 20 and 30 mg. L⁻¹, respectively*

Approximately 98% of the Cd added to serum bottles was removed after the 1st incubation (Fig. 5.2B). But the Cd concentration in the medium gradually increased with each successive transfer. After the 15th transfer, the Cd concentration in the medium became stable and ~73% of Cd remained in the aqueous phase until the end of the enrichment (Fig. 5.2B). After centrifugation, Cd in the pellet (phase II) and Cd (and/or as colloidal CdSe) in the supernatant (phase III) were measured. Fig. 5.2B shows that both kept increasing with time and became stable after the 16th successive transfer. After the 1st incubation, the Cd concentration in the pellet and supernatant was 0.2 and 0.1 mg. L⁻¹, respectively (Fig. 5.2B). While, at the end of the enrichment (300 days), Cd in the pellet and supernatant was 6.9 and 15.1 mg. L⁻¹, respectively.

5.3.2. Optical properties and size distribution and of the selenium nanoparticles

TEM images and DLS revealed that the size of the particles present in the supernatant (phase III) was smaller with a more narrow size distribution (Fig. 5.3). The biogenic CdSe particles were spherical in shape (Fig. 5.3) and had sizes ranging from 10 to 40 nm with an average size of 15.7 nm (Fig. 5.3). Particles in the pellet (phase II) were, however, larger in size ranging from 42 to 190 nm with an average size of 69 nm (data not shown).

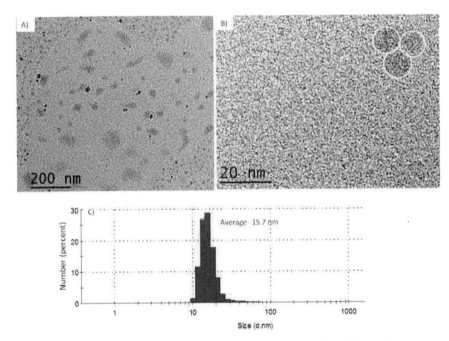

*Fig. 5.3: Transmission electron microscopic images of the CdSe NPs in the supernatant (phase III): (**A and B**) TEM images obtained at different magnifications, **C)** Particles size distribution of the CdSe NPs measured by dynamic light scattering*

The UV-Vis absorption and fluorescence spectra of both the pellet and the supernatant were recorded to determine their optical properties (Fig. 5.4). The UV-Vis absorption spectra (280 - 580 nm) of the supernatant aliquots showed the presence of absorbance bands centered at 350 nm along with a strong absorption band around 270 - 280 nm (Fig. 5.4A). However, no absorbance band was observed in the pellet sample around 350 nm and the absorbance intensity near 280 nm was also very low. The fluorescence emission spectrum was monitored using a constant excitation wavelength of 365 nm (Fig. 5.4B). The peak emission wavelength was around 450 nm in both cases with a very wide full width at half maximum (fwhm). But it is evident from Fig. 5.4B that the supernatant (phase III) shows a very strong fluorescent intensity, while the pellet (phase II) obtained from the aqueous phase showed negligible fluorescence intensity.

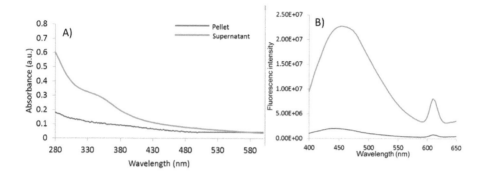

*Fig. 5.4 **A)** UV-visible spectra and **B)** fluorescence spectra of supernatant (phase III) containing CdSe and pellet (phase II) containing mainly Se(0)*

5.3.3. Selenium speciation and CdSe NPs characterization

Sludge (phase I), pellet (phase II) and supernatant (phase III) (Fig. 5.1) were further analysed by XPS and Raman spectroscopy. The XPS spectra show the presence of Cd, Se, C, O, N, and Na in the samples (Table 5.1). Intense C1s peaks were in the spectra at 284.7, 285.1, 285.7, 287.8 and 288.8 eV. The O1s peaks appear either at 531.7 or 532.4 eV. The Cd $3d_{5/2}$ and $3d_{3/2}$ peaks occurred at a binding energy of around 405 eV and 412 eV, respectively in the sludge and supernatant sample (Fig. 5.5). A weak Cd signal was observed in the pellet obtained from the aqueous phase. Three distinct Se peaks were visible at different binding energies (Fig. 5.5). An intense spectrum of the Se 3d peak appears at 56.1 and 55.5 eV, respectively, in the sludge and pellet sample. The Se 3d peak in the supernatant sample was observed at 54.9 eV (Fig. 5.5).

All three samples (phase I to III) exhibited a strong peak around 238 cm^{-1}. Another weak band around 142 - 145 cm^{-1}was also observed in all three samples. The Raman spectrum of the sludge (phase I) and supernatant (phase III) recorded another band at about 201 and 206 cm^{-1}, respectively (Fig. 5.6). However, the intensity of the peak was much higher in the supernatant sample than in the sludge sample. Interestingly, the recorded Raman spectra show another peak at 296 cm^{-1}, which was only evident in the sludge sample (Fig. 5.6).

5.4. Discussion

5.4.1. Selenite bioreduction by anaerobic granular sludge in presence of cadmium

This study demonstrated the enrichment of a microbial community in anaerobic granular sludge capable of reducing selenite to selenide in the presence of Cd coupled to the formation of CdSe NPs. Anaerobic granular sludge was enriched for > 300 days in the presence of Cd and Se(IV) for the removal and recovery of Se in the form of CdSe NPs (Fig. 5.2). During this enrichment, the increase in Cd concentration to 50 mg. L^{-1} resulted in an inhibition effect on the selenite reduction (Fig. 5.2A). At the same time, the Cd concentration in the aqueous phase also increased (Fig. 5.2B). The decrease in Cd removal with every successive transfer might be due to the saturation of binding sites for biosorption by granular sludge (King et al., 2008; Yang et al., 2010). The presence of higher Cd concentrations in the aqueous phase can generate toxic effects on microorganisms (Bartacek et al., 2008). This possibly leads to the inhibition of selenite bioreduction and eventually lesser Se(0) and Se(-II)synthesis (Fig. 5.2A) (Guo et al., 2010; Mal et al., 2016b).

Roux et al. (2001) reported that cadmium-resistant *Ralstonia metallidurans* CH34 can only reduce Se(IV) to Se(0), but not up to Se(-II). Fellowes et al. (2013) showed that *Veillonella atypica* can reduce Se(IV) up to Se(-II). The supernatant of selenite reducing *V. atypica* culture was then used for synthesizing CdSe NPs by adding $CdCl_2O_8$ separately. Recently, Ayano et al. (2013, 2014) isolated a cadmium resistant *Pseudomonas* sp. strain RB from a soil sample which was able to reduce selenite up to selenide even in the presence of 180 mg. L^{-1} Cd. All these studies show that cadmium tolerant bacteria capable of reducing selenite up to selenide are not ubiquitous in the environment, thus necessitating enrichment strategies. Moreover, there are only a few studies which focused on the microbial reduction of selenium oxyanions beyond elemental selenium up to selenide (Nancharaiah & Lens, 2015).

The results obtained in the present study show that a methanogenic consortium was capable of selenite reduction up to selenide can be enriched when incubating anaerobic granular sludge in the presence of Cd, which is a requisite for synthesizing CdSe NPs. However, complete inhibition of methanogenesis was reported in the presence of 1 mM of Se oxyanions (Lenz et al., 2011), while anaerobic microbial processes (e.g. dissimilatory reduction), detoxification mechanisms, Se-Cd interactions and other processes (e.g. adsorption and precipitation) contributes to removal of Se from selenite-containing medium (Mal et al., 2016a; Nancharaiah & Lens, 2015). Repeated exposure of methanogenic anaerobic granular sludge to selenite and Cd(II) for 300 days may have caused enrichment of selenite-reducing bacteria.The concentration of Se(0) and Se(-II) increased with time (Fig. 5.2A), also suggesting that a selenium reducing microbial community in the granular sludge was indeed enriched in the presence of the Cd and Se(IV).

Similar to Cd, it is likely that the selenite removal mechanisms, particularly via biosorption were repressed after the repetitive transfers of medium resulting in an increased total Se concentration in the aqueous phase with time (Fig. 5.2A) (Mal et al., 2016b). Thus, more selenite was available for bioreduction by the microorganisms in the aqueous phase, which finally converted to relatively higher amounts of Se(0) and Se(-II) in the aqueous phase. Moreover, due to the enrichment of selenium reducing microorganisms, selenite reduction gradually increased and complete Se(IV) reduction was observed without any Cd inhibition, resulting in a higher generation of Se(0) and Se(-II) (Fig. 5.2A).

Fig. 5.2B shows that with the increase in Se(0) in the aqueous phase, more Cd was removed along with Se(0) by centrifugation. Most possibly Cd was adsorbed onto Se(0) NPs and precipitated in the pellet probably as Se(0)-Cd complexes (Jain et al., 2016; Yuan et al., 2016). Interestingly, speciation analysis using Visual MINTEQ simulation suggested that all the Se(-II) reacts with the free metal ion to form metal selenides when the aqueous heavy metal concentration equals or exceeds the available Se(-II) (Mal et al., 2016b). Hence, the increase in the Se(-II) concentration in the aqueous phase suggests the potential reaction between the free Cd^{2+} ions and Se(-II) leading to the formation of CdSe colloids (Fig. 5.2). Depending on the size and surface properties, CdSe can remain in the aqueous phase or can precipitate along with the Se(0)-Cd complex. These results also explain that the increased Cd concentration in the aqueous phase with time is not only as free Cd ions, but as a mix of the Se(0)-Cd complex, CdSe and free Cd ions (Fig. 5.2).

5.4.2. Selenium speciation and CdSe characterization in aqueous phase

Aqueous supernatant (phase III) obtained after high-speed centrifugation showed an absorbance band around 350 nm, which is an indication for the presence of colloidal CdSe NPs (Fig. 5.4A) (Kumar et al., 2007; Suresh, 2014). In contrast, the re-suspended pellet (phase

II) did not show obvious absorption peaks for CdSe NPs (Fig. 5.4B). Absorption in the UV region up to 280 nm (Fig. 5.4A) also indicated the presence of aromatic rings of amino acids and proteins, which probably acted as a capping agent of the CdSe and also the Se(0) NPs (Kumar et al., 2007; Suresh, 2014). Control reactions in the absence of granular sludge showed no absorbance band at 350 nm, clearly implying the role of microorganisms in selenite reduction and CdSe NPs synthesis.

Photoluminescence further confirmed the formation and presence of biogenic CdSe NPs in the supernatant (phase III), but not in the pellet (phase II) (Fig. 5.4B). An emission band centered at 453 nm was observed upon excitation at 365 nm (Fig. 5.4B) mainly in the supernatant, which is attributable to the band gap or near band gap emission resulting from the recombination of the electron–hole pairs in the CdSe nanoparticles. This observation was comparable to the emission peaks of CdSe nanoparticles synthesized by chemical and biological methods reported earlier (Firth et al., 2004; Kumar et al., 2007; Suresh, 2014). Chemical analysis showed the presence of both Se and Cd elements in the pellet (phase II), thus possibilities exists for the presence of Se(0) NPs, Se(0)-Cd complex and/or a mix of Se(0) and larger CdSe NPs with no optical signatures. Particle size distributions obtained from the re-suspended pellet (42 - 190 nm) and the supernatant (10 - 40 nm) were in agreement with the observed fluorescence properties (Fig. 5.3).

XPS spectra of pellet and supernatant obtained from the aqueous phase also supported these results (Fig. 5.5). Distinct signals for Se 3d appeared at 55.5 eV for pellet (phase II), suggesting the presence of elemental Se (Han et al., 2013; Yuan et al., 2016). The Se 3d peak for CdSe generally occurred at a binding energy of 54.8 eV (Kumar et al., 2007; Suresh, 2014). The Se 3d peak in the supernatant (phase III) occurred at a binding energy of 54.9 eV suggesting the formation of CdSe (Fig. 5.5). The core level spectrum of Cd3d in the supernatant sample shows that the peaks at ~405 and ~ 412 eV (Fig. 5.5) ascribable to Cd3d$_{5/2}$ and Cd3d$_{3/2}$ agree with the core level binding energies of CdSe reported earlier (Bhande et al., 2015; Deepaa et al., 2010; Subila et al., 2013). The Cd peaks are due to CdSe and not due to CdO is further confirmed from the fact that the corresponding O1s appeared at ~ 532 eV (Table 5.1) and not in the range of 528 - 531 eV, which is normally associated with metal oxide (Deepaa et al., 2010). Consequently, the Cd3d peak can be unambiguously assigned to CdSe.

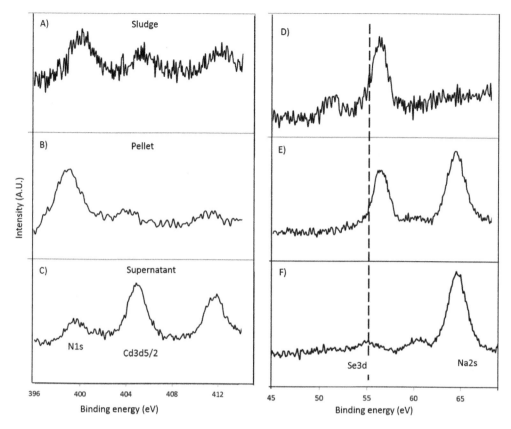

Fig. 5.5 X-ray photoelectron spectroscopic measurements of granular sludge (phase I), pellet (phase II), and supernatant (phase III). Panels (A - C) and (D – F) are the core level spectra of Cd 3d and Se 3d, respectively

In order to further confirm the formation and to evaluate the sample purity and composition of the Se and/CdSe NPs, the Cd:Se ratio of all three phases was also determined by XPS. In the supernatant (phase III), it the Cd:Se ratio was ~ 3:1. Suresh, (2014) reported a similar surface composition of biogenic CdSe QDs when the Cd:Se ratio was 3.28:0.6. XPS data was also compared with quantitative analysis during enrichment of granular sludge in the presence of selenite and Cd. From Fig. 5.2, it was revealed that the average ratio of Cd:Se was ~ 5:1 in the supernantat, while in pellet the Cd:Se ratio was 1:8. However, Cd could not be detected in the pellet (phase I) due to low signal to noise ratio during XPS analysis. To have better understanding on Cd-Se interaction and sample purity further studies (i.e. X-ray absorption spectroscopy (XAS)) are required which were out of the scope of this study. (Fellowes et al., 2013; Pearce et al., 2008).

The C1s core level spectrum of CdSe (Table 5.1) shows a signal at ~ 285, 287.7, 288.6 eV due to C-C/C-H, C=O and C-O structures, respectively (Bhande et al., 2015; Deepaa et al., 2010; Suresh, 2014; Yuan et al., 2016). The binding energy for N 1s was centred at ~ 399.5 eV (Fig. 5.5), which is possibly attributed to the nitrogen containing groups such as amine and amide groups (Yuan et al., 2016). These data further confirm that the particles were capped by the extracellular polymeric substances (EPS), which again corroborated with the observed optical properties (Fig. 5.4A).

Raman spectroscopy was chosen for further characterization as it can unambiguously distinguish the presence of elemental selenium and CdSe (Dzhagan et al., 2013; Dzhagan et al., 2007; Torchynsk & Vorobiev, 2011; Zou & Weaver, 1999). Supernatant (phase III) showed a distinct peak at 206 cm^{-1}(Fig. 6) which can be attributed to the longitudinal optical (LO) phonon mode of CdSe (Dzhagan et al., 2013; Torchynska et al., 2008). The downward red-shift of the LO peak from its bulk CdSe value at 210 - 213 cm^{-1}due to phonon confinement is reported in other studies (Meulenberg et al., 2004; Tanaka et al., 1992). The main limitation of the microbial synthesis process described in this study, however, was the presence of trigonal Se (t-Se) in both phases, suggesting that the CdSe nanoparticles need to be separated and purified from the Se nanoparticles. Both the pellet and supernatant phases show a peak at 238 cm^{-1} (Fig. 5.6), corresponding to the vibrational mode of –Se–Se–Se– chains (Iovu et al., 2005; Kotkata et al., 2009). This particular peak of t-Se was the most intense indicating the presence of a large amount of t-Se as Se defects (Fig. 5.6) along with the CdSe NPs.

For the economic production of metal selenide nanoparticles, it is imperative that the process should be scaled up on an industrial scale. UASB reactors have already been used successfully for treating several types of wastewaters at industrial scale. Recent studies have shown that anaerobic granular sludge based reactors are suitable for removal andrecovery of critical elements like tellurium and palladium (Mal et al., 2017; Pat-Espadas et al., 2016; Pat-Espadas et al., 2015; Ramos-Ruiz et al., 2016). The results presented here are useful for the development of a selenium oxyanion-reducing bioreactor to simultaneously remove selenium oxyanions and cadmium from wastewaters by combining bioremediation of environmentally toxic selenium-rich wastewater with biorecovery of Se as CdSe NPs. Selenium principally exists in the form of oxyanions (selenite or selenate) in wastewaters. An important aspect of this study is thus the use of selenite as precursor, which is more environmentally relevant and is different from other selenium compounds like SeCl$_4$ used in previous studies (Kumar et al., 2007; Suresh, 2014).

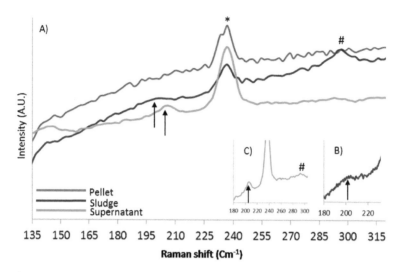

*Fig. 5.5 Raman spectra of sludge (phase I), pellet (phase II) and supernatant (phase III). All three phases contain mainly t-Se (*). The sludge and supernatant sample show peaks for CdSe (arrow). A weak band for CdS (#) is also visible in the sludge sample, which forms a CdSe/CdS core/shell structure. Inset: peak for CdSe (arrow) in sludge (**B**) and in supernatant (**C**)*

5.4.3.   Characterization of biomass associated CdSe NPs

XPS spectra (phase I) show that the Se 3d peak of the sludge sample is broad (Fig. 5.5), thus making it impossible to separate these two peaks from elemental Se and CdSe. Hence, XPS alone is not sufficient in this case as the chemical shift of Se or Cd is very low in CdSe. The core level spectrum of Cd3d in the sludge sample, however, clearly shows that the peaks at ~405 and ~ 412 eV (Fig. 5.5) similar to the supernatant sample ascribable to Cd3d$_{5/2}$ and Cd3d$_{3/2}$ agree with the core level binding energies of CdSe reported earlier (Bhande et al., 2015; Deepaa et al., 2010; Subila et al., 2013). In sludge (phase I), the Cd:Se ratio was around 1:4, which is in agreement with Fig 5.2. The average Cd:Se ratio of the deposition onto the sludge was 1:3.8.

Table 5.1: Binding energy (eV) of the three phases obtained from the XPS analysis

Sample (Charging)	C1s	Cd 3d5/2	Se 3d	Na1s	N1s	O1S
Sludge (phase I)	285.1	405.5	56.1	Nil	400.3	532.4
(ΔE 3.3 eV)			52.0			
Pellet (phase II)	285.1	405.1	55.9	63.3	399.5	531.9
(ΔE 4.7eV)	288.6					
Supernatant (phase III)	284.7	405.1	55.0	62.9	399.5	531.7
(ΔE 3.4eV)	287.8		51.5			

Raman spectroscopy of the sludge (phase I) sample shows a strong peak at 238 cm^{-1} (Fig. 5.6), corresponding to the vibrational mode of –Se–Se–Se– chains indicating the deposition of large amounts of t-Se on the granular sludge (Iovu et al., 2005; Kotkata et al., 2009). It further depicts that the local oscillator (LO) frequency of CdSe red shifted further to 201 cm^{-1}(Fig. 5.6), probably due to the decrease in CdSe thickness or due to the formation of a CdSe/CdS core/shell structure (Dzhagan et al., 2013; Torchynska et al., 2008). This suggests that both the crystalline CdSe core and Se defects are present simultaneously, regardless of their association to the sludge or presence in aqueous phase as colloidal form. Fig. 5.6 shows another peak at 296 cm^{-1} in the Raman spectrum for the sludge sample, which is attributed to the Cd-S vibration related peak (Dzhagan et al., 2013). Downward shifting of the CdS-like peak from its bulk value of the LO phonon at 305 cm^{-1} indicate formation of an alloyed layer of CdS_xSe_{1-x} at the interface between the CdSe core and CdS shell (Dinger et al., 1999; Dzhagan et al., 2013). For nanoparticles with a core/shell structure with 2 and 3 nm thick shells, the dominant mode in this spectral region is evident between 290 and 300 cm^{-1}and is assigned to the CdS LO phonon of the CdS shell (Dzhagan et al., 2013).

The Cd ions might interact with the sulfhydryl (-SH) group mainly present in the EPS originating from anaerobic granular sludge biofilm, leading to the formation of a CdS shell structure around a CdSe core. A similar interaction between Cd(II) and -SH occurred during the biosynthesis of CdS NPs using EPS extracted from *Pseudomonas aeruginosa* (Raj et al., 2016). Mak et al. (2011) reported that when CdTe QDs were capped with different capping agents like 3-mercaptopropionic acid, thioglycolic acid or thioglycerol thiol groups, a similar interaction between the -SH group of the capping agent and the core of the QDs and formation of CdS$_x$Te$_{1-x}$ was observed. The absence of a proper CdS LO phonon peak in the supernatant sample (Fig. 5.6) was probably due to the partial alloying with the core (Dzhagan et al., 2013).

Based on the data, a possible mechanism for the formation of biogenic CdSe NPs under the experimental conditions is proposed (Fig. 5.7). The proposed mechanism of the formation of Se nanostructures by enriched anaerobic granular sludge includes the following steps: 1) microbial reduction of selenite to Se(0) and then to Se(-II), 2) interaction of Cd^{2+} with sulphydryl groups (-SH) of EPS to form [Cd-S-EPS]$^{2+}$, 3) interaction between [Cd-S-EPS]$^{2+}$ and Se(0) to form a Cd-Se(0)-EPS complex, and 4) interaction between [Cd-EPS]$^{2+}$ and Se(-II) to form CdSe-EPS (Fig. 5.7). The formation of CdSe could be both in the aqueous phase as well as in the granular sludge depending on their size and surface properties. The probability of Cd interacting with –SH group of EPS in the aqueous phase is much less than in the sludge sample. As a result, the real thickness of the CdS shell could be less and not enough to show a proper peak for the CdS shell (Fig. 5.6).

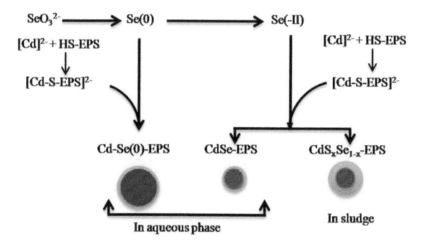

Fig. 5.7 Proposed mechanism of CdSe nanostrucutre formation during microbial reduction of selenite in the presence of Cd(II). HS-EPS: sulphydryl groups (-SH) of EPS

The implications of the present study thus go beyond biosynthesis of CdSe NPs. Cadmium can be present in different selenium containing waste streams including acid mine drainage and flue gas desulfurization waters as co-contaminant with selenium oxyanions (Mal et al., 2016b). It is clear that different Se compounds such as Se(0), Se(0)-Cd complex, CdSe NPs and CdS_xSe_{1-x} are formed when selenite is bioreduced in the presence of Cd. This is important because deposition of Se(0)-Cd complex, CdSe NPs and even CdS_xSe_{1-x}may form an important fraction of the total Se during bioremediation of selenium and cadmium-rich wastewater or at a site that undergoes *in-situ* remediation. Furthermore, they may pose hazard to the ecosystems in which they are released, especially to the organisms required to maintain ecosystem homeostasis, such as bacteria, fungal populations, and plants (Mal et al., 2016a; Wiecinski et al., 2013). Exposure of the CdSe core to an oxidative environment can cause decomposition and desorption of Cd ions and Se ions, which may plays an important role in subsequent toxicity (Mal et al., 2016a). Engineered NPs cannot be considered as a uniform group of substances. Hence further studies are required as different forms of selenide NPs (e.g. CdSe and CdS_xSe_{1-x}) may have different physcio-chemical properties (e.g. dissolution, uptake) in different environmental condition than that of Se(0) and can influence the efficiency of the total remediation process.

5.5. Conclusions

This work demonstrated the enrichment of anaerobic granular sludge in the presence of Cd(II) and selenite over a period of 10 months and the feasibility of coupling bioremediation of Se-containing wastewater and recovery of selenium as CdSe NPs. The microbial community was successfully enriched to reduce selenite up to selenide in the presence of Cd. Optical properties of the aqueous phase confirmed the formation of CdSe NPs. XPS and Raman spectroscopy data showed that CdSe was present either as entrapped in the EPS or as colloidal NPs in the aqueous phase. Particularly, the presence of extracellular CdSe NPs in the aqueous phase was highly significant and favorable as these allow an easier separation and recovery step. The major limitation, however, was the presence of t-Se along with CdSe, suggesting that a better separation and purification method for harvesting the CdSe NPs is required.

Acknowledgements

This research was supported through the Erasmus Mundus Joint Doctorate Environmental Technologies for Contaminated Solids, Soils, and Sediments (ETeCoS[3]) (FPA n[0] 2010-0009) and the BioMatch project No. 103922 (Role of biofilm-matrix components in the extracellular reduction and recovery of chalcogen) funded by the European Commission Marie Curie International Incoming Fellowship (MC-IIF). The authors would like to thank Dr. Chloé

Fourdrin (Université Paris-Est Marne-la-Vallée) for the technical help for RAMAN analysis and Dr. Santanu Bera (Bhabha Atomic Research Centre, Kalpakkam, India) for XPS analysis.

References

Ayano, H., Kuroda, M., Soda, S., Ike, M. 2014. Effects of culture conditions of Pseudomonas aeruginosa strain RB on the synthesis of CdSe nanoparticles. *J Biosci Bioeng.*, **119**(4), 440-445.

Ayano, H., Miyake, M., Terasawa, K., Kuroda, M., Soda, S., Sakaguchi, T., Ike, M. 2013. Isolation of a selenitereducing and cadmium-resistant bacterium Pseudomonas sp. strain RB for microbial synthesis of CdSe nanoparticles. *J Biosci Bioeng.*, **117**(5), 576-581.

Bartacek, J., Fermoso, F.G., Baldó-Urrutia, A.M., van Hullebusch, E.D., Lens, P.N.L. 2008. Cobalt toxicity in anaerobic granular sludge: influence of chemical speciation. *J Ind Microbiol Biotechnol.*, **35**(11), 1465-1474.

Bhande, S.S., Ambade, R.B., Shinde, D.V., Ambade, S.B., Patil, S.A., Naushad, M., Mane, R.S., Alothman, Z.A., Lee, S.H., Han, S.H. 2015. Improved photoelectrochemical cell performance of tin oxide with functionalized-multiwalled carbon nanotubes-cadmium selenide sensitizer. *ACS Appl Mater Interfaces.*, **7**(45), 25094-25104.

Boyd, R. 2011. Selenium stories. *Nat Chem.*, **3**, 570.

Deepaa, M., Gakhar, R., Joshi, A.G., Singh, B.P., Srivastav, A.K. 2010. Enhanced photoelectrochemistry and interactions in cadmium selenide-functionalized multiwalled carbon nanotube composite films. *Electrochimica Acta.*, **55**(22), 6731-6742.

Dessì, P., Jain, R., Singh, S., Seder-Colomina, M., van Hullebusch, E.D., Rene, E.R., Ahammad, S.Z., Lens, P.N.L. 2016. Effect of temperature on selenium removal from wastewater by uasb reactors. *Water Res.*, **94**, 146-154.

Dinger, A., Hetterich, M., Goppert, M., Grun, M., Weise, B., Liang, J., Wagner, V., Geurts, J. 1999. Growth of CdS/ZnS strained layer superlattices on GaAs(0 0 1) by molecular-beam epitaxy with special reference to their structural properties and lattice dynamics. *J Cryst Growth.*, **200**, 391-398.

Dzhagan, V., Valakh, M., Milekhin, A., Yeryukov, N., Zahn, D., Cassette, E., Pons, T., Dubertret, B. 2013. Raman- and IR-active phonons in CdSe/CdS core/shell nanocrystals in the presence of interface alloying and strain. *J Phys Chem. C*, **117**, 18225-18233.

Dzhagan, V.M., Valakh, M.Y., Raevskaya, A.E., Stroyuk, A.L., Kuchmiy, S.Y., Zahn, D.R.T. 2007. Resonant Raman scattering study of CdSe nanocrystals passivated with CdS and ZnS. *Nanotechnology.*, **18**, 285701-285707.

Fellowes, J.W., Pattrick, R.A.D., Lloyd, J.R., Charnock, J.M., Coker, V.S., Mosselmans, W., Weng, T.C., Pearce, C.I. 2013. Ex situ formation of metal selenide quantum dots using bacterially derived selenide precursors. *Nanotechnology*, **24**(14), 145603-145612.

Firth, A.V., Haggata, S.W., Khanna, P.K., Williams, S.J., Allen, J.W., Magennis, S.W., Samuel, I.D.W., Cole-Hamilton, D.J. 2004. Production and luminescent properties of CdSe and CdS nanoparticlepolymer composite. *J Liminesc.*, **109**, 163-172.

Guo, H., Luo, S., Chen, L., Xiao, X., Xi, Q., Wei, W., Zeng, G., He, Y. 2010. Bioremediation of heavy metals by growing hyperaccumulaor endophytic bacterium *Bacillus* sp. L14. *Bioresour Technol.*, **101**, 8599-8605.

Han, D.S., Batchelor, B., Abdel-Waha, A. 2013. XPS Analysis of sorption of selenium(IV) and selenium(VI) to mackinawite (FeS). *Environ Prog Sustain Energy.*, **32**, 84-93.

Iovu, M., Kamitsos, E., Varsamis, C., Boolchand, P., Popescu, M. 2005. Raman spectra of AsxSe100-x glasses doped with metals. *Chalcogenide Lett.*, **2**(3), 21-25.

Jain, R., Dominic, D., Jordan, N., E.R., R., Weiss, S., E.D., v.H., Hübner, R., P.N.L., L. 2016. Higher Cd adsorption on biogenic elemental selenium nanoparticles. *Environ Chem Lett.*

King, P., Rakesh, N., Lahari, B.S., Kumar, P.Y., Prasad, V.S.R.K. 2008. Biosorption of zinc onto *Syzygium cumini* L.: Equilibrium and kinetic studies. *Chem Eng J.*, **144**(2), 181-187.

Kotkata, M.F., Masoud, A.E., Mohamed, M.B., Mahmoud, E.A. 2009. Structural characterization of chemically synthesized CdSe nanoparticles. *Physica E.*, **41**, 640-645.

Kumar, S.A., Ansary, A.A., Ahmad, A., Khan, M.I. 2007. Extracellular biosynthesis of CdSe Quantum Dots by the fungus, *Fusarium Oxysporum. J Biomed Nanotechnol.*, **3**(2), 190-194.

Mak, J.C.W., Farah, A.A., Chen, F., A.S., H. 2011. Photonic crystal fiber for efficient raman scattering of CdTe quantum dots in aqueous solution. *ACS Nano*, **5**(5), 3823-3830.

Mal, J., Nancharaiah, Y.V., Maheshwari, N., van Hullebusch, E.D., Lens, P.N. 2017. Continuous removal and recovery of tellurium in an upflow anaerobic granular sludge bed reactor. *J Hazard Mater.*, **327**, 79-88.

Mal, J., Nancharaiah, Y.V., van Hullebusch, E.D., Lens, P.N.L. 2016a. Metal Chalcogenide quantum dots: biotechnological synthesis and applications. *RSC Adv.*, **6**, 41477-41495.

Mal, J., Nancharaiah, Y.V., van Hullebusch, E.D., Lens, P.N.L. 2016b. Effect of heavy metal co contaminants on selenite bioreduction by anaerobic granular sludge. *Bioresour Technol.*, **206**, 1-8.

Meulenberg, R.W., Jennings, T., Stroue, G.F. 2004. Compressive and tensile stress in colloidal CdSe semiconductor quantum dots. *Phys Rev B.*, **70**, 235311-235321.

Nancharaiah, Y.V., Lens, P.N.L. 2015. Ecology and biotechnology of selenium-respiring bacteria. *Microbiol Mol Biol Rev.*, **79**(1), 61-80.

Pat-Espadas, A.M., Field, J.A., Otero-Gonzalez, L., Razo-Flores, E., Cervantes, F.J., Sierra-Alvarez, R. 2016. Recovery of palladium(II) by methanogenic granular sludge. *Chemosphere.*, **144**, 745-573.

Pat-Espadas, A.M., Field, J.A., Razo-Flores, E., Cervantes, F.J., Sierra-Alvarez, R. 2015. Continuous removal and recovery of palladium in an upflow anaerobic granular sludge bed (UASB) reactor. *J Chem Technol Biotechnol.*, **91**(4), 1183–1189.

Pawar, V., Kumar, R.A., Zinjarde, S., Gosavi, S. 2013. Bioinspired inimitable cadmium telluride quantum dots for bioimaging purposes. *J Nanosci Nanotechnol.*, **13**(6), 3826-3831.

Pearce, C.I., Coker, V.S., Charnock, J.M., Pattrick, R.A.D., Mosselmans, J.F.W., Law, N., Lloyd, R. 2008. Microbial manufacture of chalcogenide-based nanoparticles via the reduction of selenite using *Veillonella atypica* : an in situ EXAFS study. *Nanotechnology*, **19**(5), 156603-156615.

Raj, R., Dalei, K., Chakraborty, J., Das, S. 2016. Extracellular polymeric substances of a marine bacterium mediated synthesis of CdS nanoparticles for removal of cadmium from aqueous solution. *J Colloid Interface Sci.*, **462**, 166-175.

Ramos-Ruiz, A., Field, J.A., Wilkening, J.V., Sierra-Alvarez, R. 2016. Recovery of elemental tellurium nanoparticles by the reduction of tellurium oxyanions in a methanogenic microbial consortium. *Environ Sci Technol.*, **50**(3), 1492-500.

Roest, K., Heilig, H.G.H.J., Smidt, H., Vos de, W.M., Stams, A.J.M., Akkermans, A.D.L. 2005. Community analysis of a full-scale anaerobic bioreactor treating paper mill wastewater. *Syst Appl Microbiol.*, **28**, 175-185.

Roux, M., Sarret, G., Pignot-Paintrand, I., Fontecave, V., Coves, J. 2001. Mobilization of selenite by *Ralstonia metallidurans* CH34. *Appl Environ Microbiol.*, **67**, 769-773.

Subila, K.B., Kumar, G.K., Shivaprasad, S.M., Thomas, K.G. 2013. Luminescence properties of CdSe quantum dots: Role of crystal structure and surface composition. *J Phys Chem Lett.*, **4**, 2774-2779.

Suresh, A.K. 2014. Extracellular bio-production and characterization of small monodispersed CdSe quantum dot nanocrystallites. *Spectrochim Acta Mol Biomol Spectrosc.*, **130**, 344-349.

Tanaka, A., Onari, S., Arai, T. 1992. Raman scattering from CdSe microcrystals embedded in a geramante glass matrix. *Phys Rev B.*, **45**(12), 6587-6592.

Torchynsk, A.T., Vorobiev, Y. 2011. Semiconductor II-VI Quantum Dots with Interface States and Their Biomedical Applications.

Torchynska, T.V., Douda, J., Ostapenko, S., Jimenez-Sandoval, S., Phelan, C., Zajac, A., Zhukov, T., Sellers, T. 2008. Raman scattering study in bio-conjugated core-shell CdSe/ZnS quantum dots. *J Non-Crystal Solids.*, **354**, 2885-2890.

Wiecinski, P.N., Metz, K.M., King Heiden, T.C., Louis, K.M., Mangham, A.N., Hamers, R.J., Heideman, W., Pedersen, J.A. 2013. Toxicity of oxidatively degraded quantum dots to developing zebrafish (*Danio rerio*). *Environ Sci Technol.*, **47**(16), 9132–9139.

Yan, Z., Qian, J., Gu, Y., Su, Y., Ai, X., Wu, S. 2014. Green biosynthesis of biocompatible CdSe quantum dots in living *Escherichia coli* cells. *Mater Res Exp.*, **1**, 15401-15415.

Yang, C., Wang, J., Lei, M., Xie, G., Zeng, G., Luo, S. 2010. Biosorption of zinc(II) from aqueous solution by dried activated sludge. *J Environ Sci.*, **22**(5), 675-680.

Yuan, F., Song, C., Sun, X., Tan, L., Wanga, Y., Wang, S. 2016. Adsorption of Cd(II) from aqueous solution by biogenic selenium nanoparticle. *RSC Adv.*, **6**, 15201-15209.

Zou, S., Weaver, M.J. 1999. Surface-enhanced Raman spectroscopy of cadmium sulfidercadmium selenide superlattices formed on gold by electrochemical atomic-layer epitaxy. *Chem Phys Lett.*, **321**, 101-107.

CHAPTER 6

Modification of extracellular polymeric substances (EPS) of anaerobic granular sludge used for synthesis of cadmium selenide nanoparticles

Abstract

This study investigated the compositional changes in the extracellular polymeric substances (EPS) matrix of anaerobic granular sludge enriched for synthesizing cadmium selenide (CdSe) nanoparticles. The methanogenic anaerobic granular sludge was enriched in the presence of cadmium (10-50 mg L^{-1}) and selenite (79 mg L^{-1}) for 300 days at pH 7.3 and 30 °C in a fed-batch experiment. *In vitro* experiments on metal(loid)–EPS interactions showed that CdSe nanoparticles found mainly in the loosely bound-EPS. Analysis of the EPS composition revealed a large increase in protein content (3 times) and a decrease in humic-like substances content (0.5 times) in the enriched granular sludge (EGS) compared to untreated granular sludge. EPS fingerprints, obtained by size exclusion chromatography (SEC) coupled to a fluorescence detector, showed a significant increase in the intensity of protein-like substances of >100 kDa aMW (apparent molecular weight) in the EPS matrix of EGS. This was accompanied by a prominent decrease in protein-like substances of aMW <10 kDa. The fingerprint of humic-like substances showed emergence of a new peak with aMW of 13 to 300 kDa in the EPS extracted from EGS. These results clearly indicate a compositional change in the EPS matrix of EGS synthesizing CdSe nanoparticles.

Keywords: Extracellular polymeric substances (EPS), Size exclusion chromatography (SEC) fingerprint, Molecule size distribution, Anaerobic granular sludge, Selenium

6.1. Introduction

Selenium (Se) is a contaminant of potential concern in natural environments primarily present in acid mine drainage, acid seeps and agricultural drainage (Mal et al., 2016a). Microbial reduction of Se-oxyanions (i.e. selenate and selenite) to elemental selenium (Se(0)) has emerged as a leading technology for bioremediation and treatment of Se-bearing wastewaters (Mal et al., 2016b; Nancharaiah & Lens, 2015; van Hullebusch, 2017). However, metal ions such as cadmium or zinc present in the selenium-rich wastewaters influence the microbial reduction and the fate of biogenic Se(0) (Mal et al., 2016a). When metal ions are available, it can either form metal – Se(0) complexes due to adsorption of metals onto Se(0) nanoparticles (NPs) or metal-selenides NPs after reacting with selenide (HSe^-) generated during the microbial reduction of Se-oxyanions (Jain et al., 2016; Mal et al., 2016a). For example, microbial reduction of selenite (Se(IV)) in the presence of Cd(II) by anaerobic granular sludge results in the formation and deposition of Se(0) NPs, Se(0)-Cd complexes and/or cadmium selenide (CdSe) NPs by the reaction of Cd(II) with biogenic HSe^- (Mal et al., 2017a).

Extracellular polymeric substances (EPS) are a complex mixture of high molecular weight macromolecules comprising mainly of proteins, carbohydrates and humic-like substances (Bhatia et al., 2013; Flemming et al., 2007). The content and composition of EPS from aerobic or anaerobic microbial aggregates play a crucial role in the functioning of biological wastewater treatment processes (Bhatia et al., 2013; Mal et al., 2017b). The EPS also influence the bioflocculation and sludge settleability (Li & Yang, 2007). Being the outer layer of microbial cells, EPS make contact and interact with the metals and nanoparticles before entering cells and thus play a vital role in protecting the microbial cells (Li & Yu, 2014; Li & Yang, 2007; Sheng et al., 2010; Tang et al., 2017). The EPS matrix is known to play a potential role in microbe-metal interactions, i.e. biosorption, bioreduction and biomineralization, thus determining the biogeochemical cycling of metal ions as well as the transformation and immobilization of metals in natural environments (Li et al., 2016; Taylor et al., 2016; Tourney & Ngwenya, 2014; Vindedahl et al., 2016).

Previous studies have shown that EPS of granular sludge play a pivotal role in governing the surface characteristics and fate of biogenic Se(0) NPs (Jain et al., 2015). Raj et al. (2016) have shown that EPS extracted from *Pseudomonas aeruginosa* JP-11 isolated from marine environment facilitated stabilization of cadmium sulphide (CdS) NPs formed via chemical reduction. Recently, we have reported biological synthesis of CdSe/CdS core/shell structure along with CdSe NPs using anaerobic granular sludge. This was possible by the interaction of Cd(II) ions available on the surface of CdSe NPs with the sulfhydryl (–SH) groups of the EPS components (Mal et al., 2017a). It was hypothesized that EPS play a prominent role in binding of Cd(II), Se(IV) and thus act as nucleation site for the formation of CdSe NPs. However, to the best of our knowledge, there is no detailed study on compositional changes in the EPS matrix of biofilms engaged in reduction of Se oxyanions and formation of NPs. Therefore, to better engineer metal biosorption or microbial synthesis of CdSe, it is essential to have an in-depth understanding of the composition and fingerprint of EPS as well as the metal(loid) – EPS interactions.

The main objective of this research was, thus, to investigate the differences in the EPS extracted from anaerobic granular sludge enriched in the presence of Cd(II) and Se(IV) for microbial synthesis of CdSe NPs. The EPS matrix was extracted from untreated granular sludge (UGS) and after enriching anaerobic granular sludge (EGS) for CdSe biosynthesis upon prolonged exposure to Cd(II) and Se(IV) ions for detailed characterization. Further experiments were performed on metal(loid) – EPS interactions *in vitro* to decipher the role of EPS in CdSe biosynthesis. The three-dimensional excitation-emission matrix (3D-EMM) of the EPS was used to understand the changes in chemical properties of fluorescence compounds in the EPS. Size exclusion chromatography (SEC) coupled to fluorescence detectors was used to differentiate protein and humic-like substances fingerprints in the EPS extracted from anaerobic granules before and after the enrichment to determine the compositional changes in the EPS matrix due to enrichment of granular sludge in the presence of Cd(II) and Se(IV).

6.2. Materials and methods

6.2.1. Source of biomass and enrichment of granular sludge in Cd(II)/Se(IV) medium

Anaerobic granular sludge was collected from a full scale upflow anaerobic sludge blanket (UASB) reactor treating paper mill wastewater (Industriewater Eerbeek B.V., Eerbeek, The Netherlands). A detailed description of the anaerobic granular sludge was given in detail by Roest et al. (2005). Additional characterization of the anaerobic granular sludge was performed with compression tests of granular sludge.

The microbial community of the anaerobic granular sludge was enriched for 300 days in the presence of Cd(II) and Se(IV) at pH 7.3 and 30 °C in a fed-batch experiment in order to produce CdSe NPs (Mal et al., 2017a). Details of the enrichment of the anaerobic granular sludge, synthesis of Se(0) and CdSe NPs, separation and characterization of Se(0) and CdSe NPs are reported in a previous paper.(Mal et al., 2017a) The initial Cd(II) concentration was increased in a stepwise manner from 10 to 50 mg L^{-1} and then later decreased to 30 mg L^{-1} as Se(IV) reduction was negatively affected and was not complete in the presence of 50 mg L^{-1} of Cd(II). Sodium lactate (5 mM) and sodium selenite (1 mM = 79 mg L^{-1}) were used, respectively, as the carbon and selenite source (Mal et al., 2017a). Se(0) NPs were collected by centrifugation at 20,000×g for 20 min at 4°C used for characterization of the EPS associated with it as a capping agent.

A LLOYD FR K plus Ametek instrument was used to detect "breaks/compressions" with a LC 5 Newton probe (LLOYD Ametek). Rupture 2 software was used for data treatment. The Newton/mm (N mm^{-1}) coefficients were determined to characterize the compressive strength of the granules. 13 replicates were performed for untreated anaerobic granular sludge (UGS) and Se/Cd enriched anaerobic granular sludge (EGS). Determination of compressive strengths show higher values from enriched granules (11.5 ± 0.8 N mm^{-1}) compared to untreated granules (8 ± 1 N mm^{-1}).

6.2.2. EPS extraction and characterization

For detailed characterization of EPS and to investigate the differences in EPS fingerprints, EPS was extracted from UGS and EGS according to a previous protocol with modification (D'Abzac et al., 2010a). Prior to extraction, granular sludge was washed twice with Milli-Q water. UGS and EGS were centrifuged at 10,000×g for 20 min at 4°C and the supernatants were collected as loosely bound EPS (LB-EPS) extracts. The pellet was re-suspended in Milli-Q water and used for extracting tightly bound EPS (TB-EPS). Subsequent heating at 60 °C for 10 min followed by centrifugation at 10,000×g for 20 min at 4°C was performed for extracting TB-EPS from the same granular sludge samples.

The biochemical compositions, i.e. humic-like (HS), protein (PN) (using Lowry modified by Frølund et al. 1995) and polysaccharide (PS) (using phenol-sulfuric acid method by Dubois et al. 1956) substances, of the EPS extracts was determined as described earlier (D'Abzac et al., 2010a). The total organic carbon (TOC), total Se and Cd content in the EPS extracts was quantified using Phoenix 8000 TOC-analyzer (Dohrmann) and a ICP-MS (Agilent 7700x), respectively.

6.2.3. Excitation Emission Matrix (EEM) spectra of the EPS by fluorescence spectroscopy

Excitation Emission Matrix (EEM) spectra of the EPS extracts were obtained at 22 ($\pm$ 1) °C by using a spectro-fluorophotometer. Emission was scanned from 220-500 nm after excitation ranging from 220-400 nm using 10 nm increments (Fig. 6.1). The fluorescence data were processed using the Panorama Fluorescence 3.1 software (LabCognition, Japan). When needed, extracts were diluted with 50 mM phosphate buffer (25 mM Na_2HPO_4 + 25 mM NaH_2PO_4) at pH 7.0 ($\pm$ 0.1) (Bhatia et al., 2013).

6.2.4. SEC with diode array and fluorescence detector

Fingerprints of EPS extracts from UGS and EGS were obtained as described previously Bhatia et al. (2013) with a Merck Hitachi LA Chrom chromatograph equipped with a L7200 autosampler, a L7100 quaternary pump, a D7000 interface and a L7485 fluorescence detector. To improve the resolution of EPS fingerprints, two columns were used in series (Bourven et al., 2015): Bio SEC 300Å and 100Å (Agilent), with theoretical selective permeation ranges of 5 - 1,250 kDa and 0.1 - 100 kDa, respectively. 50 mM phosphate buffer (25 mM Na_2HPO_4 and 25 mM NaH_2PO_4, Prolabo) with 150mM NaCl at pH 7.0 ($\pm$ 0.1) was used as mobile phase at a constant flow of 0.7 mL/min. Mass calibration was performed for both protein and humic-like substances. For proteins, the following molecules were used: Ferritine (Sigma) - 440 kDa; Immunoglobulin G from human serum (Sigma) - 155 kDa; Bovine Serum Albumin (Sigma) - 69 kDa; Ribonuclease A (Sigma) - 13.7 kDa and Thyrotropin releasing hormone (Sigma) - 362.38 Da (Bourven et al., 2015). Humic-like substances calibration curves was made with synthetic polymers of polystyrene sulfonate (Sigma): 350 kDa, 150 kDa, 77 kDa, 32 kDa, 13 kDa, 4.3 kDa, 510 Da.

The logarithm of the MW (log (MW)) was plotted as a function of the elution volume (Ve).

Calibration with protein standards:

$Ve = -3.35 \, LogMW + 31.05$ \qquad $(R^2 = 0.9805)$ \qquad Eq. (1)

Calibration with polystyrene sulfonate standards:

$Ve = -5.87 \, LogMW + 41.17$ \qquad $(R^2 = 0.9638)$ \qquad Eq. (2)

where, MW is the molecular weight in Da and Ve represents the elution volume in mL. The fingerprints were monitored by fluorescence detection. Excitation/emission wavelengths determined from EEM spectra were: 221/350 nm for protein-like substances and 345/443 nm for humic-like substances. Signals were standardized as a function of the TOC content of the EPS sample to allow comparison of the chromatograms.

A 100 µL of filtered (0.2 µm, Whatman™, GE Health) EPS extracts (LB-EPS and TB-EPS) was injected for each analysis. In order to remove potential metal ions retained in the columns, a cleaning step was applied after each chromatographic run (Polec-Pawlak et al., 2005). Se(0) NPs samples were also analysed by SEC after filtering with a 0.2 µm membrane to characterize the EPS associated with Se(0) NPs. Fractions were collected during SEC analysis for each EPS sample at different elution times to study metal-EPS interactions: the five collected fractions are detailed in Fig. 6.2A. Se and Cd concentrations in each fraction were measured by ICP-MS (Agilent 7700x). These concentrations were used to determine the repartition (% of Se or Cd) in a given fraction compared to the sum of element detected in each fraction.

6.2.5.   In vitro experiments on metal(loid) – EPS interactions

In control experiments, 0.1 mM each of Cd(II) and Se(IV) was added to the LB-EPS extract extracted from UGS. EEM spectra were collected after 20 min of incubation to study the metal(loid)-EPS interaction and fluorescence quenching. In order to study a potential impact of the presence of Cd and Se on SEC-Fluorescence analysis, the same LB-EPS extracts incubated with both Cd(II) and Se(IV) were also analysed by SEC and compared with the fingerprints of LB-EPS extracted from EGS.

6.3. Result

6.3.1. EPS biochemical composition

Table 6.1 summarizes the characteristics and the main biochemical composition of the EPS extracted from UGS and EGS. There was no significant difference in total LB-EPS as measured by TOC after the enrichment of sludge, but the total extracted TB-EPS was ~6 times lower. PN content of LB and TB EPS were three and six times higher, respectively for EGS (Table 6.1). HS-like and PS contents were standardized to PN content to highlight a strong decrease of HS-like after Cd/Se enrichment especially in TB-EPS (10-fold reduction). The PS/PN ratios also decreased but to a lower extent (2-fold reduction). In contrast, the standardized HS-like and PS contents were similar in LB and TB-EPS for UGS (Table 6.1). Se(0) NPs also contained significant amount of organic matter. PN content is lower than in EPS extracted from EGS (10 and 6 times lower compared to LB and TB-EPS, respectively). The standardized PS content is the same as in EGS EPS but the standardized HS-like is 4 times higher than in EGS EPS.

Table 6.1 Main characteristics of LB and TB-EPS samples extracted from UGS and EGS (Number of replicate; N= 2)

	LB-EPS		TB-EPS		Se-EPS
	UGS	EGS	UGS	EGS	
TOC (mgC. L^{-1})	15 ± 1	17 ± 1	724 ± 9	126 ± 6	75 ± 5
Biochemical composition of EPS (mg. mgC^{-1} of EPS)					
Protein	2.03*	6.4*	0.47*	3.64 ± 0.06	0.59*
Humic substance	2.31*	1.4 ± 0.24	1.04*	0.56*	0.46*
Polysaccharide	1.4 ± 0.05	1.87 ± 0.06	0.33*	1.00*	0.20*
Se and Cd concentration (µg. mgC^{-1} of EPS)					
Total Se	-	70.6 ± 0.7	-	27.6 ± 0.1	7983 ± 37
Total Cd	-	45.8 ± 0.9	0.55*	10.10 ± 0.2	472 ± 11.5

UGS – Untreated anaerobic granular sludge; EGS – Enriched anaerobic granular sludge

* - Negligible standard deviation (< 0.05)

The Se and Cd concentrations measured in the EPS extracts are given in Table 6.1. Selenium was not detected in the EPS of UGS before enrichment, while only a small amount of Cd (0.11 µg mg C^{-1} of EPS) was found in the TB-EPS. The total Se and Cd concentration was found to be 70.6 and 45.8 µg mg C^{-1} of EPS, respectively in the LB-EPS extracted from the EGS, while it was only 27.6 and 10.1 µg mg C^{-1} of EPS, respectively in the TB-EPS extracted. The Se(0) NPs separated from the aqueous phase by centrifugation contained ~ 8 mg Se L^{-1} and 0.5 mg Cd L^{-1} (Table 6.1).

6.3.2. Fluorescence properties of the EPS extracts

EEM spectra corresponding to different EPS extracts are presented in Fig 6.1. The 3-D spectra were divided into five different regions (Bhatia et al., 2013; Chen et al., 2003). Each region is associated with different compounds: derived from PN – tyrosine and tryptophan (I or II) corresponding to aromatic proteins, fulvic-like acids (III), soluble microbial by-product-like (SMP) substances (IV) and HS-like substances (V) (Bhatia et al., 2013; Chen et al., 2003). All the EPS extracts (Fig 6.1) exhibited fluorescence in the areas associated with the PN - tyrosine and tryptophan (I or II) at excitation/emission wavelengths (Ex/Em) of 220-250/275-375 nm. The relative intensity of fluorescence of PN was higher in both LB and TB-EPS (Fig 6.1B and D) extracts after the enrichment. The fluorescence of humic-like substances at Ex/Em wavelengths of 310-350/275-375 nm was only observed in LB-EPS in UGS before enrichment (Fig 6.1C). These observations were used to determine the optimal settings for fluorescence detection for SEC. The following excitation/emission wavelengths were thus used: 221/350 nm for generating protein-like fingerprints and 345/443 nm for humic-like fingerprints.

6.3.3. Fingerprints of EPS extract by SEC

The SEC chromatograms of EPS extracted from UGS and EGS are given in Fig 6.2. The fingerprints of PN-like molecules for LB-EPS from untreated and enriched sludge (Fig 6.2A) are quite similar: they only differ by the intensity of the chromatographic peaks. The same observation can be made for TB-EPS (Fig 6.2B). Both LB and TB-EPS fingerprints show an increase in intensity in the range of higher aMW of >100 kDa after the enrichment in the presence of Cd(II) and Se(IV). In contrast, the presence of lower aMW <10 kDa PN-like substances became negligible after the enrichment, particularly in LB-EPS.

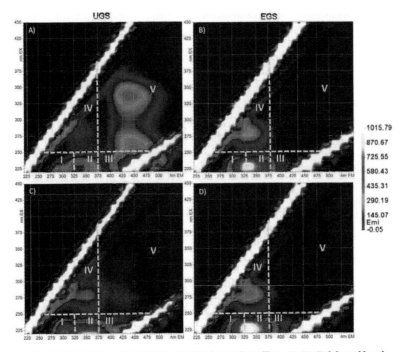

I, II: aromatic proteins – IV : soluble microbial by-products-like – III, V : Fulvic and humic-like substances

Fig. 6.1 Fluorescence EEM (Excitation Emission Matrix) of extracted EPS in phosphate buffer at pH 7.0 (± 0.1): LB-EPS (A and B) and TB-EPS (C and D) for untreated (UGS) and exposed (EGS) granular sludge (dilution factor: 50x (A and B), 1000x (C), 500x (D))

For the UGS, the HS-like fingerprints are quite similar for LB and TB-EPS (Fig. 6.2C and 6.2D), three main fractions with aMW ≤ 13 kDa are observed, but the fluorescence intensities are about 10-fold higher for LB-EPS. After Cd and Se enrichment, both LB and TB-EPS fingerprints are modified and differ one from the other. For LB-EPS, compared to the UGS, intensities are strongly reduced (~20 fold lower) and the first peak is eluted significantly later, corresponding to a decrease of the aMW. TB-EPS fingerprints are quite similar to LB-EPS after enrichment at the exception of the apparition of a broad peak corresponding to aMW in the range 13 to 300 kDa.

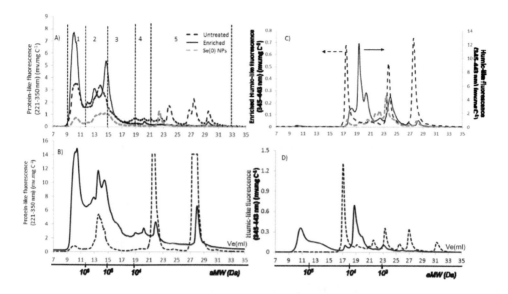

Fig. 6.2 SEC chromatograms of LB (A, C) and TB-EPS (B, D) from UGS and Cd/Se enriched EGS and EPS associated with Se(0) NPs. Fluorescence detection at Ex/Em: 221/350 nm for protein-like substances (A, B) and at 345/443 nm for humic-like substances (C, D). Vertical dot lines in (A) correspond to the five fractions collected for determining Cd and Se concentrations in LB-EPS extracted from EGS. Legend symbols are inside the panel (A)

The SEC chromatograms of Se(0) NPs are given in Fig 6.2 and are compared with LB EPS extracted from UGS and EGS. The fingerprints of PN-like substances (Fig 6.2A) are quite similar but about two to five fold lower peaks intensities for EPS associated with Se(0) NPs compared to LB-EPS from UGS and EGS, respectively. Interestingly, for HS-like substances (Fig 6.2C), Se(0) NPs fingerprint is quite similar to LB-EPS from UGS, but not EGS with about 20-fold lower peaks intensities.

6.3.4. Metal(loid)-EPS interactions

In a control (*in-vitro*) experiment, significant fluorescence quenching was evident after the addition of Cd(II) and Se(IV) to the LB-EPS extracted from UGS (Fig 6.3). It is clear that fluorescence intensities for fulvic and humic-like substances (III and V) decreased, while it became negligible for aromatic protein-like (I and II) substances after the addition of Se(IV) (alone or with Cd) (Fig 6.3). Se(IV) thus induces a specific fluorescence quenching on regions attributed to PN and HS-like substances. LB-EPS of UGS samples were analysed further by

SEC before and after spiking with Cd and Se(IV) to study the effect of Cd and Se ions on its elution profile (Fig 6.4). Fig 6.4 reveals that addition of Cd or Se(IV) to the LB-EPS did not cause any significant changes in the fingerprint of PN and HS-like molecules for LB-EPS.

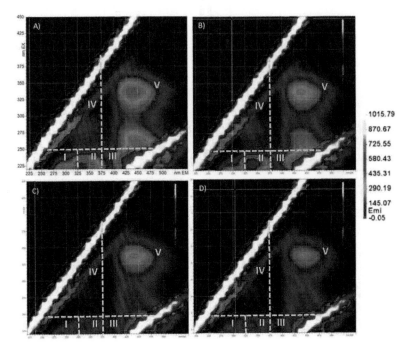

Fig. 6.3 Fluorescence EEM of a 50-fold diluted solution in phosphate buffer at pH 7.0 (± 0.1) of LB-EPS extracted from UGS before (A) and after addition of 0.1 mM Cd (B), 0.1 mM Se(IV) (C) or 0.1 mM Cd and 0.1 mM Se(IV) (D). Incubation time: 20 mins

6.3.5.   Fractionation of metal(loid)s associated with the extracted EPS

In order to investigate the metal(loid)s, i.e. Cd(II) and Se(IV) interaction with the EPS components, five different fractions were collected during the SEC analysis of EPS extracted from EGS (fractions are indicated in Fig 6.2A). The Cd and Se repartition in each fraction is given in Table 6.2. Se was only detected in the 4th fraction (19 – 21 mL), which is close to the total permeation volume, and thus corresponds to the elution of very small compounds. In LB-EPS, Cd was also mainly found in the 4th fraction (89% of total Cd), whereas in TB-EPS the majority of Cd (63%) was present in the 1st fraction (9 – 12 mL). Se(0) NPs (size <200 nm) were also injected and it was found that Se was eluted at the 4th fraction and majority of Cd was mainly present in the 1st fraction and 5th fraction (21-33 mL) (Table 6.2).

*Table 6.2 Metal(loid)s (Cd and Se) fractionations in EPS samples extracted from EGS. (L.D. – below low detection and N.D. – No detection), * - Negligible standard deviation (< 0.05)*

Fractions		1	2	3	4	5
Elution volumne		(9-12 mL)	(12-15 mL)	(15-19 mL)	(19-21 mL)	(21-33 mL)
LB-EPS	Se	<LD	<LD	<LD	74	<LD
	Cd	0.88	0.26	0.28	35.77	3.09
TB-EPS	Se	<LD	<LD	<LD	43.51	<LD
	Cd	8.55	1.19	1.19	0.88	1.73
Se NPs	Se	<LD	<LD	<LD	507.95	<LD
	Cd	0.86	0.4	0.24	0.25	1.34

Limit of Detection: Se (1 μg L^{-1}), Cd (0.05 μg L^{-1})

6.4. Discussion

6.4.1. Effect of enrichment on EPS quantity and granular strength

Results The sludge compression test shows that both density and compressibility increased after enrichment of granular sludge in presence Cd and Se(IV) for 300 days. Li and Yang (2007) reported that higher LB-EPS was correlated with poor flocculation and settleability of activated sludge. No significant differences were noticed in the LB-EPS content of UGS and EGS (Table 6.1), which could either be due to i) a lower content in EPS or ii) a less efficient EPS extraction (Zhang et al., 1999). However, the decrease of EPS content of the enriched granules in the present study is not consistent with the statements of the literature which claim that metal stress (at sub-lethal concentration) generally induces a higher EPS content (Sheng et al., 2010). The lower extraction efficiency could be a result of the higher granular strength as highlighted by compression tests. The compositional changes i.e. higher PN and PS, lower HS-like substances (Table 6.1) can also have a beneficial role in improved density and compressibility of granular sludge. These observations are in agreement with a previous study on the role of EPS components on physical properties of activated sludge collected from different full-scale wastewater treatment plants which treated domestic and industrial waste. Sludge flocculation increased with an increase in PN and PS content of sludge, but decreased with an increase in HS-like substances (Wilen et al., 2003).

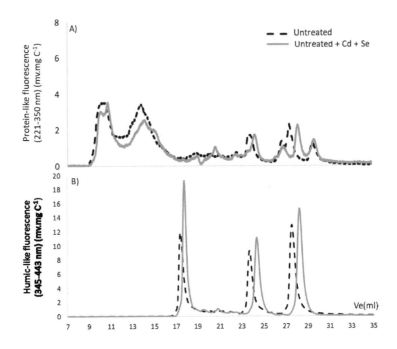

Fig. 6.4 SEC chromatograms of LB-EPS extracted from UGS before and after addition of 0.1 mM Cd and 0.1 mM Se(IV). Fluorescence detection at Ex/Em: 221/350 nm for protein-like substances (A) and at 345/443 nm for humic-like substances (B)

Metal(loid) cations (e.g., Cd(II), Pb(II) or Se(IV)) tend to complex with the components of the EPS matrix. These ion bridging interactions between metal(loid) cations and the EPS components increase the flocculation of activated sludge (Sheng et al., 2010). Significant fluorescence quenching was observed when Cd and Se(IV) were added to the LB-EPS extracted from UGS (Fig 6.3), which clearly indicates that LB-EPS has a strong binding capacity for Cd(II) and Se(IV). The presence of high concentrations of Cd and Se in the EPS extracts of EGS (Table 6.1) also support the argument and indicate strong metal(loid)-EPS interactions. This implicates the involvement of some of the organic ligands emanating fluorescence upon complexation of EPS with Cd(II) and Se(IV). It is assumed that these metal(loid) – EPS interactions may play a beneficial role in sludge flocculation and compressibility.

6.4.2.   Changes in biochemical composition of EPS extracts

PN, PS and HS-like substances are the major components of the EPS matrix which influence the physicochemical properties of sludge in biological wastewater treatment systems (Bhatia et al., 2013). The composition and role of PN and PS and their formation dynamics in a particular system are of great importance (Sheng et al., 2010). Proteins play a vital role in maintaining the structure, properties and functions of microbial aggregates (Wang et al., 2005; Zhu et al., 2015). Both LB and TB-EPS extracted from the EGS had a higher PN/PS ratio. This could be due to the stress exerted by the metal(loid)s or their NPs, i.e. Cd and Se, Se(0) or CdSe NPs on the microorganisms. Some studies have reported that extracellular proteins are more important than polysaccharides in mediating electrostatic bonds in the biofilm matrix due to the presence of negatively charged amino acids in the proteins (Cao et al., 2011), thus have a high metal binding affinity (Pan et al., 2010).

Moreover, proteins play a crucial role in microbial synthesis of Se(0) NPs including transport of Se oxanions (in)to the cell, the redox reactions, export of Se(0) nuclei out of the cell and assembly of Se(0) into NPs (Tugarova & Kamnev, 2017). Thus, an increase of existing protein content and/or apparition of new specific proteins in the EPS matrix of the granular sludge might offer a better biosorption for Cd(II) and Se(IV), redox reactions and capping of Se(0) and CdSe NPs in microbial synthesis of Se(0) and CdSe NPs. It should be noted that HS-like substances are the second major component of Se(0) NPs, unlike EPS (Table 6.1), thus the HS-like substances of Se(0) NPs could also contribute to nucleation for CdSe formation. This hypothesis is consistent with the quenching of fluorescence of HS-like substance when the EPS are incubated with Se (Fig 6.3). The HS-like substances were not originated from the synthetic growth media and could thus come from the granule itself and/or be a by-product of the bacterial metabolism.

6.4.3.   Fingerprint of EPS using SEC

SEC revealed a distinct aMW distribution for PN and HS-like substances in the EPS extracted from sludge exposed to Cd and Se(IV) compared to the unexposed biomass (Fig 6.2). The intensity of the high molecular weight (HMW) of >100 kDa PN-like substances increased in the EPS after the enrichment, while the low molecular weight (LMW) of <10 kDa PN-like substances disappeared. The proteins with HMW offer more binding sites and interaction points with the metal cations and polymers, which offer a higher structural stability to the granules (Zhu et al., 2015). On the other hand, LMW substances are generally considered as building blocks and low-molecular weight acids (< 1 kDa). Building blocks are generally comprised of low-molar-mass organic acids (including amino acids, peptides), while the low amphiphilic (slightly hydrophobic) compounds represent sugars, alcohols, aldehydes and ketones. So, possible conversion of these LMW substances (amino acids and peptides) to HMW biopolymers during repetitive exposure of UGS to Cd(II) and Se(IV) for 300 days cannot be ruled out (Aryal et al., 2009).

However, LMW substances can also interact with Cd(II) and Se(IV) and form bigger molecules that are eluted as HMW substances. In order to further confirm the changes in EPS composition, a LB-EPS of UGS sample was analysed before and after spiking with Cd and Se(IV) to study the effect of Cd and Se ions on its elution profile. Addition of Cd or Se(IV) to the LB-EPS did not cause any significant changes in the elution profile of the LB-EPS (Fig 6.4). As a result, the changes in fingerprint of PN- or HS-like substances revealed previously can be attributed to modifications in EPS composition due to long exposure of granular sludge to Cd and Se(IV), and not to a simple alteration of fluorescence detector response due to metal(loid) addition to EPS. These changes of EPS composition in enriched granules, could thus be due to a modification of i) metabolism and/or ii) microbial communities after the enrichment, and/or iii) the release of different EPS components, i.e. PN and PS. Further studies (e.g. microbial communities, proteomics, Cd and Se interaction with HS-like substances) should be carried out to improve the details of the EPS composition (e.g. different functional groups of EPS and changes in binding properties, identification of proteins associated with the Se(0) or CdSe NPs produced by the anaerobic granular sludge) and their role in microbial synthesis of nanoparticles.

Recently Gonzalez-Gil et al. (2016) also indicated that changes in abundance of proteins in the outer layer of granular sludge takes place during the microbial synthesis of Se(0) from Se(IV) by anaerobic granular sludge. The majority of proteins associated to the Se(0) NPs were ascribed to the γ-Proteobacteria family *Pseudomonadaceae,* and specifically to the genus *Pseudomonas* which was totally contrasting to the protein profile of the extracellular matrix of the selenite reducing granular sludge, in which the majority of the proteins were affiliated to methanogenic *Archaea* and δ-*Proteobacteria.* Concerning HS-like substances, the EPS mass distribution for UGS is consistent with previous studies (Bhatia et al., 2013): aMW are close to 10 kDa or lower. In contrast, exposure to Se and Cd induced the appearance of HS-like substances of TB-EPS with high aMW (13 to 300 kDA) (Fig 6.2D). This could be due to the association of HS with others compounds (other HS or proteins) favoured by the bridging effect of the divalent cation Cd with a similar role as Ca (Sobeck & Higgins, 2002).

Concerning Se(0) NPs, surprisingly, PN-like substances which represent the major part of Se(0) NPs's organic matter, have a similar aMW repartition to LB EPS. This similarity can lead to the assumption that the nucleation of Se(0) NPs partly involves PN-like substances present in LB-EPS. For HS-like substances, the fingerprint obtained for Se(0) NPs and LB-EPS of EGS is rather unique (Fig 6.2C), with some common peaks with LB-EPS from UGS, which suggest that part of HS-like origin of Se(0) NPs comes from metabolized LB-EPS HS-like substance.

6.4.4.   Interaction of Se and Cd with EPS

Large quantities of Se and Cd were found in the EPS after the enrichment of granular sludge in the presence of Cd and Se (Table 6.1). This can be due to the biosorption of Cd and Se(IV) and/or precipitation of Se(0) or CdSe NP, formed during the bioreduction of selenite in the presence of Cd (Mal et al., 2016a). It was also suggested that Se(0) and CdSe NPs or Te(0) NPs formed after microbial reduction of selenite or tellurite was associated mainly with the LB-EPS fraction of the granular sludge (Mal et al., 2017a; Mal et al., 2017b). The total Se and Cd concentrations were 2.6 and 4.5 times higher, respectively in the LB-EPS than the TB-EPS extracted from EGS (Table 6.1) further confirming that majority of Cd and Se are retained in the LB-EPS, but not in the TB-EPS.

Moreover, Se and Cd eluted separately in the TB-EPS possibly as Se(0) and Cd-EPS complex but not as CdSe NPs (Table 6.2). Recently, we found that Cd strongly interact with the –SH of EPS and forms an [Cd–S–EPS] complex (Mal et al., 2017a). This might result in the formation of large aggregates of HS-substance-Cd complexes or association of Cd with macromolecules in TB-EPS. It can also explain the appearance of a big peak of HS-like substances of TB-EPS with aMW > 100 kDa, but not in the LB-EPS (Fig 6.2). The majority of Cd eluted at the same, i.e. (1st) fraction (9 – 12 mL), also corroborated this finding (Table 6.2). In contrast, the majority of Se and Cd elute together in the 4th fraction (19 – 21 mL) (Table 6.2) in LB-EPS, corresponding to the total permeation volume indicating that either both Se and Cd were present in ionic form or were present as smaller NPs (i.e. CdSe).

However, there was no selenite found in the enriched sludge and selenium was only present in the sludge, either as Se(0), CdSe or CdS_xSe_{1-x} with a size ranging from 10-190 nm (Mal et al., 2017a). Thus, it is possible that due to the formation of CdSe and CdS_xSe_{1-x} NPs with very small size Cd and Se elute together at the total permeation volume. Interestingly, when Se(0) NPs with a size of <200 nm were run through SEC, the majority of the Se eluted similarly in the 4th fraction (Table 6.2). All these results suggest that Se(0) and CdSe NPs formation found mainly in the LB-EPS, highlighting the pivotal role of LB-EPS in microbial synthesis of Se(0) and CdSe NPs by providing the microbial reduction and nucleation sites for NPs synthesis. Thus, the influence of EPS in microbial synthesis of CdSe should not be ignored.

6.5. Conclusion

Size-exclusion chromatography fingerprints were used to investigate the compositional changes in EPS after the enrichment of anaerobic granular sludge in the presence of Cd(II) and Se(IV) for microbial synthesis of CdSe. The PN and PS-like content increased, while the concentration of HS-like substances decreased and variations in mass distribution for both PN and HS-like substances were observed. HMW protein-like substances of >100 kDa increased significantly, whereas LMW protein-like substances of <10 kDa disappeared. The differences observed in the fingerprints of HS-like substances were not only in the peak intensities but also in peak positions, meaning that the composition of the EPS changes in terms of concentration and nature of compounds. Total metal(loid) analysis and fractionation by SEC of Se/Cd enriched sludge revealed that the majority of Se and Cd was present mainly in the LB-EPS and eluted together, possibly as CdS_xSe_{1-x} or CdSe NPs.

Acknowledgements

This research was supported through the Erasmus Mundus Joint Doctorate Environmental Technologies for Contaminated Solids, Soils, and Sediments (ETeCoS3) (FPA n00 2010-0009) and the European Commission Marie Curie International Incoming Fellowship (MC-IIF) Role of biofilm-matrix components in the extracellular reduction and recovery of chalcogens (BioMatch project, No. 103922). The authors also acknowledge the funding from the Earth System Science and Environmental Management (ESSEM) COST Action ES1302 European Network on Ecological Functions of Trace Metals in Anaerobic Biotechnologies to support a Short-term scientific mission (STSM) at the Groupement de Recherche Eau Sol Environnement, Université de Limoges, France.

References

Aryal, R., Lebegue, J., Vigneswaran, S., Kandasamy, J., Grasmick, A. 2009. Identification and characterisation of biofilm formed on membrane bio-reactor. *Sep Purif Technol.*, **67**, 86–94.

Bhatia, D., Bourven, I., Simon, S., Bordas, F., van Hullebusch, E.D., Rossano, S., Lens, P.N.L., Guibaud, G. 2013. Fluorescence detection to determine proteins and humic-like substances fingerprints ofexopolymeric substances (EPS) from biological sludges performed bysize exclusion chromatography(SEC). *Bioresour Technol.*, **131**, 159-165.

Bourven, I., Simon, S., Bhatia, D., van Hullebusch, E.D., Guibaud, G. 2015. Effect of various Size Exclusion Chromatography (SEC) columns on the fingerprints of Extracellular Polymeric Substanc (EPS) extracted from biological sludge. *J Taiwan Inst Chem Engg.*, **49**, 148-155.

Cao, B., Shi, L., Brown, R.N., Xiong, Y., Fredrickson, J.K., Romine, M.F., Marshall, M.J., Lipton, M.S., Beyenal, H. 2011. Extracellular polymeric substances from *Shewanella* sp. HRCR-1 biofilms: characterization by infrared spectroscopy and proteomics. *Environ Microbiol.*, **13**(4), 1018-1031.

Chen, W., Westerhoff, P., Leenheer, J.A., Booksh, K. 2003. Fluorescence excitation-emission matrix regional integration to quantify spectra for dissolved organic matter. *Environ Sci Technol.*, **37**(24), 5701-510.

D'Abzac, P., Bordas, F., van Hullebusch, E.D., Lens, P.N.L., Guibaud, G. 2010. Extraction of extracellular polymeric substances (EPS) from anaerobic granular sludges: comparison of chemical and physical extraction protocols. *Appl Microbiol Biotechnol.*, **85**(5), 1589-1599.

Flemming, H.C., Neu, T.R., Wozniak, D.J. 2007. The EPS matrix: The "House of Biofilm Cells". *. J Bacteriol.*, **189**, 7945-7947.

Gonzalez-Gil, G., Lens, P.N.L., Saikaly, P. 2016. Selenite reduction by anaerobic microbial aggregates: Microbial community structure, and proteins associated to the produced selenium spheres. *Front Microbiol.*, **7**, 571-598.

Jain, R., Dominic, D., Jordan, N., Rene, E.R., Weiss, S., van Hullebusch, E.D., Hübner, R., Lens, P.N.L. 2016. Higher Cd adsorption on biogenic elemental selenium nanoparticles. *Environ Chem Lett.* .

Jain, R., Jordan, N., Weiss, S., Foerstendorf, H., Heim, K., Kacker, R., Hübner, R., Kramer, H., van Hullebusch, E.D., Lens, P.N.L. 2015. Extracellular polymeric substances govern the surface charge of biogenic elemental selenium nanoparticles. *Environ Sci Technol.*, **49**(3), 1713-1720.

Li, S.W., Zhang, X., Sheng, G.P. 2016. Silver nanoparticles formation by extracellular polymeric substances (EPS) from electroactive bacter. *Environ Sci Pollut Res Int.*, **23**(9), 8627-8633.

Li, W.W., Yu, H.Q. 2014. Insight into the roles of microbial extracellular polymer substances in metal biosorption. *Bioresour Technol.*, **160**, 15-23.

Li, X.Y., Yang, S.F. 2007. Influence of loosely bound extracellular polymeric substances (EPS) on the flocculation, sedimentation and dewaterability of activated sludge. *Water Res.*, **41**(5), 1022-1030.

Mal, J., Nancharaiah, Y.V., Bera, S., Maheswari, N., van Hullebusch, E., Lens, P.N.L. 2017a. Biosynthesis of CdSe nanoparticles by anaerobic granular sludge. *Environ Sci Nano.*, **4**, 824-833.

Mal, J., Nancharaiah, Y.V., Maheswari, N., van Hullebusch, E.D., Lens, P.N.L. 2017b. Continuous removal and recovery of tellurium in an upflow anaerobic granular sludge bed reactor. *J Hazard Mater.*, **327**, 79-88.

Mal, J., Nancharaiah, Y.V., van Hullebusch, E., Lens, P.N.L. 2016b. Metal Chalcogenide quantum dots: biotechnological synthesis and applications. *RSC Adv.*, **6**, 41477-41495.

Mal, J., Nancharaiah, Y.V., van Hullebusch, E.D., Lens, P.N.L. 2016a. Effect of heavy metal co-contaminants on selenite bioreduction by anaerobic granular sludge. *Bioresour Technol.*, **206**, 1-8.

Nancharaiah, Y.V., Lens, P.N.L. 2015b. Selenium biomineralization for biotechnological applications. *Trends Biotechnol*, **33**, 323-330.

Pan, X., Liu, J., Zhang, D., Chen, X., Song, W., Wu, F. 2010. Binding of dicamba to soluble and bound extracellular polymeric substances (EPS) from aerobic activated sludge: a fluorescence quenching study. *J Colloid Interface Sci.*, **345**, 442-447.

Polec-Pawlak, K., Ruzik, R., Abramski, K., Ciurzynska, M., Gawronska, H. 2005. Cadmium speciation in *Arabidopsis thaliana* as a strategy to study metal accumulation system in plants. *Anal Chim Acta.*, **540**, 61-70.

Raj, R., Dalei, K., Chakraborty, J., Das, S. 2016. Extracellular polymeric substances of a marine bacterium mediated synthesis of CdS nanoparticles for removal of cadmium from aqueous solution *J Colloid Interface Sci.*, **462**, 166-175.

Roest, K., Heilig, H.G.H.J., Smidt, H., Stams, A.J.M., Akkermans, A.D.L. 2005. Community analysis of a full-scale anaerobic bioreactor treating paper mill wastewater. *Syst Appl Microbiol.*, **28**, 175-185.

Sheng, G.P., Yu, H.Q., Li, X.Y. 2010. Extracellular polymeric substances (EPS) of microbial aggregates in biological wastewater treatment systems, a review. *Biotechnol Adv.*, **28**, 882-894.

Sobeck, D.C., Higgins, M.J. 2002. Examination of three theories for mechanisms of cation-induced bioflocculation. *Water Res.*, **36**(3), 527-538.

Tang, J., Zhu, N., Zhu, Y., Kerrb, P., Wu, Y. 2017. Distinguishing the roles of different extracellular polymeric substance fractions of a periphytic biofilm in defending against Fe_2O_3 nanoparticle toxicity. *Environ Sci Nano.*, **4**, 1682-1691.

Taylor, C., Matzke, M., Kroll, A., Read, D.S., Svendsenb, C., Crossleya, A. 2016. Toxic interactions of different silver forms with freshwater green algae and cyanobacteria and their effects on mechanistic endpoints and the production of extracellular polymeric substances. *Environ Sci Nano.*, **2**, 396-408.

Tourney, J., Ngwenya, B.T. 2014. The role of bacterial extracellular polymeric substances in geomicrobiology. *Chem Geol.*, **386**, 115-132.

Tugarova, A.V., Kamnev, A.A. 2017. Proteins in microbial synthesis of selenium nanoparticles. *Talanta*, **174**, 539–547.

van Hullebusch, E.D. 2017. Bioremediation of Selenium contaminated wastewaters. *Springer International Publishing*, VII, 130.

Vindedahl, A.M., Strehlau, J.H., Arnold, W.A., Penn, R.L. 2016. Organic matter and iron oxide nanoparticles: aggregation, interactions, and reactivity *Environ Sci Nano.*, **3**, 494-505

Wang, Z.W., Liu, Y., Tay, J.H. 2005. Distribution of EPS and cell surface hydrophobicity in aerobic granules. *Appl Microbiol Biotechnol.*, **69**(4), 469-473.

Wilen, B.M., Jin, B., Lant, P. 2003. The influence of key chemical constituents in activated sludge on surface andflocculation properties. . *Water Res.*, **37**, 2127–2139.

Zhang, X., Bishop, P.L., Kinkle, B.K. 1999. Comparison of extraction methods for quantifying extracellular polymers in biofilms. *Wat Sci Tech.*, **39**(7), 211-218.

Zhu, L., Zhou, J., Lv, M., Yu, H., Zhao, H., Xu, X. 2015. Specific component comparison of extracellular polymeric substances (EPS) in flocs and granular sludge using EEM and SDS-PAGE. *Chemosphere.*, **126**, 26-32.

CHAPTER 7

Continuous removal and recovery of tellurium in an upflow anaerobic granular sludge bed (UASB) reactor

This chapter has been published in modified form:

Mal, J., Nancharaiah, Y.V., van Hullebusch, Maheshwari, N., Lens, P.N.L. 2017. Biosynthesis of CdSe nanoparticles by anaerobic granular sludge. Continuous removal and recovery of tellurium in an upflow anaerobic granular sludge bed reactor. *J Hazard Mater.* 327, 79-88

Abstract

Continuous removal of tellurite (TeO_3^{2-}) from synthetic wastewater and subsequent recovery in the form of elemental tellurium was studied in an upflow anaerobic granular sludge bed (UASB) reactor operated at 30°C. The UASB reactor was inoculated with anaerobic granular sludge and fed with lactate as carbon source and electron donor at an organic loading rate of 0.6 g COD L^{-1} d^{-1}. After establishing efficient and stable COD removal, the reactor was fed with 10 mg TeO_3^{2-}. L^{-1} for 42 d before increasing the influent concentration to 20 mg TeO_3^{2-}. L^{-1}. Tellurite removal (98 and 92%, respectively, from 10 and 20 mg Te. L^{-1}) was primarily mediated through bioreduction and most of the removed Te was retained in the bioreactor. Characterization using XRD, Raman spectroscopy, SEM-EDX and TEM confirmed association of tellurium with the granular sludge, typically in the form of elemental Te(0) deposits. Furthermore, application of an extracellular polymeric substances (EPS) extraction method to the tellurite reducing sludge recovered up to 78% of the tellurium retained in the granular sludge. This study demonstrates for the first time the application of a UASB reactor for continuous tellurite removal from tellurite-containing wastewater coupled to elemental Te(0) recovery.

Key words: Tellurite, Bioreduction, Te(0) recovery, anaerobic granular sludge, UASB reactor

7.1. Introduction

Tellurium (Te) is a metalloid which belongs to the chalcogen group. Like sulfur and selenium, also Te exists in four oxidation states in the environment, including Te(VI) (tellurate, TeO_4^{2-}), Te(IV) (tellurite, TeO_3^{2-}), Te(0) (elemental Te(0)), and Te(-II) (telluride, Te^{2-}) (Turner et al., 2012; Zannoni et al., 2008). Due to the excellent thermal, optical and electrical properties, Te and Te-containing compounds are extensively used in various industries such as steel and glass manufacturing, petroleum refining, solar panels, sensor production and rechargeable battery manufacturing (Ba et al., 2010; Turner et al., 2012; Zonaro et al., 2015). Te-compounds such as cadmium telluride (CdTe) quantum dots have attracted promising fluorescence applications for *in vivo* imaging and tagging cells in life sciences (Deng et al., 2007; Li et al., 2011; Turner et al., 2012). In medicine, Te compounds have traditionally been used for the treatment of bacterial infectious diseases such as leprosy, dermatitis, cystitis, and severe eye infections (Cunha et al., 2009; Zannoni et al., 2008; Zonaro et al., 2015). Other biological activities for Te-containing compounds are the inhibition of cytokine production by T cells, immunomodulatory activities and antisickling (Sredni et al., 2007).

Te is, however, one of the least abundant elements in the lithosphere (de Boer & Lammertsma, 2013; Ramos-Ruiz et al., 2016). The natural abundance of Te in the Earth's crust is only 1-5 µg/kg (Belzile & Chen, 2015), which is much lower compared to Au, Pt and rare earth elements (Belzile & Chen, 2015). Thus, development of new technologies for the recovery of Te from mining waste streams and its end-use applications are essential to ensure its sustainability. Presently, most of the Te is commercially produced as a by-product of mining and copper refining processes (Belzile & Chen, 2015). Biotechnological processes are considered as green and cost-effective strategies for removing Te from mine waste streams (or Te-associated copper ores) to avoid pollution and to recover this scarce element for meeting the tellurium supply risk (Nancharaiah et al., 2016; Ramos-Ruiz et al., 2016; Turner et al., 2012).

Microbial reduction of soluble tellurite to insoluble Te(0) has long been considered for bioremediation (Turner et al., 2012). Several bacterial cultures capable of reducing Te(IV) to Te(0) either through detoxification, redox poise maintenance or anaerobic respiration mechanisms (Borghese et al., 2016; Forootanfar et al., 2015; Kim et al., 2012; Klonowska et al., 2005; Tanaka et al., 2010) have been isolated. Baesman et al. (2007) demonstrated

Bacillus selenitireducens and *Sulfurospirillum barnesii* are able to grow using, respectively, Te(IV) and Te(VI) as electron acceptors and produce elemental Te(0) nanocrystals. Recently, Bonificio and Clarke (2014) have shown that a *Pseudomonas* sp. strain EPR3 is capable of releasing and recovering Te(0) from solid sources like copper anode slime, tellurium dioxide, CdTe and bismuth telluride. There are, however, limited studies on tellurite reduction by mixed microbial communities (i.e. biofilms or granular sludge) which are suitable for process development in bioreactors (Ramos-Ruiz et al., 2016). Moreover, tellurium recovery using anaerobic granular sludge in continuous reactors has not yet been reported.

The aim of this study was to investigate tellurite removal and recovery in an upflow anaerobic granular sludge bed (UASB) reactor inoculated with anaerobic granular sludge. The biomass associated Te(0) nanocrystals were characterized using X-ray diffraction (XRD), Raman spectroscopy, Transmission electron microscopy (TEM) and scanning electron microscopy (SEM) coupled with energy dispersive X-ray (EDX). Chemical analysis and three-dimensional excitation-emission matrix (3D-EEM) spectroscopy were used to examine the compositional changes in the extracellular polymeric substances (EPS) matrix of tellurite reducing granular sludge.

7.2. Materials and methods

7.2.1. Source of biomass and synthetic wastewater composition

Anaerobic granular sludge was collected from a full scale UASB reactor treating brewery wastewater (Biothane Systems International B.V., Delft, The Netherlands) (Gonzalez-Gil et al., 2001) and used for inoculating the UASB reactor. The composition of the synthetic wastewater was in g.L^{-1}: Na$_2$HPO$_4$.2H$_2$O (0.053), KH$_2$PO$_4$ (0.041), NH$_4$Cl (0.300), CaCl$_2$.2H$_2$O (0.010), MgCl$_2$.6H$_2$O (0.010) and NaHCO$_3$ (0.040). Lactate (3 to 5 mM) was used as the carbon source and electron donor. Trace elements were provided by adding 0.1 mL each of the acid and alkaline trace metal solutions to 1 L of the SWW as described in Stams et al. (1992). The pH of the synthetic wastewater was 7.0 - 7.1.

7.2.2. UASB reactor operation

A 1.2 L working volume polycarbonate UASB reactor (Fig. 7.1), operated as described by Lenz et al. (2008), was used for studying continuous tellurite reduction and retention of biogenic Te(0). The UASB reactor was inoculated with 200 g wet weight (DW = 12.5% w/w wet weight) of anaerobic granular sludge as described by Lenz et al. (2008). The UASB reactor was operated at 12 h hydraulic retention time (HRT) for 70 d at 30°C.

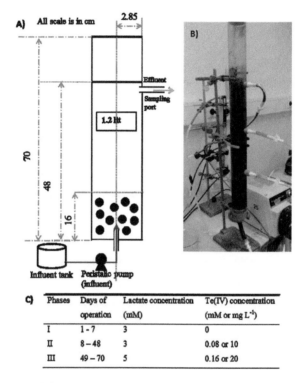

Phases	Days of operation	Lactate concentration (mM)	Te(IV) concentration (mM or mg L⁻¹)
I	1 - 7	3	0
II	8 – 48	3	0.08 or 10
III	49 – 70	5	0.16 or 20

Fig. 7.1 Configuration of the continuously operated UASB reactor (A), digital photograph of the actual UASB reactor set-up (B) used in the current study for studying tellurite reduction and Te recovery, and (C) The operating conditions of the UASB reactor used in this study

The UASB reactor was operated in three distinct periods (Fig. 7.1). The influent of the UASB reactor was supplemented with 3 mM (~ 336 mg. L⁻¹) sodium lactate (corresponding to an organic loading rate of ~0.6 g chemical oxygen demand (COD) L⁻¹ d⁻¹) as sole carbon source till the end of period II (49 days) and later increased to 5 mM (corresponding to an organic loading rate of ~0.85 g COD L⁻¹ d⁻¹) during period III until the end of the operation (days 50-70). In period 1, lasted up to day 7, the influent of the reactor was fed with solely lactate, but without addition of tellurite to start-up the reactor. In Period II (days 8-49), the influent of the

UASB reactor was supplemented with 0.08 mM potassium tellurite (K_2TeO_3) corresponding to 10 mg Te(IV). L^{-1}. In Period III (days 50-70), the Te(IV) concentration in the influent was increased to 0.16 mM K_2TeO_3 or ~20 mg Te(IV).L^{-1}. Samples from the influent and effluent of the UASB reactor were collected regularly to monitor the overall reactor performance.

7.2.3. Characterization of the Te(0) associated with the tellurite reducing granular sludge

Sludge samples were collected at the end of period III of the UASB reactor operation (day 70) for further characterization. Control and tellurite reducing granular sludge samples were analyzed using SEM-EDX. Samples were prepared by fixing them in a 2.5% glutaraldehyde phosphate-buffered saline solution (pH 7.0), subsequently dehydrated with a graded ethanol series (Ngwenya & Whiteley, 2006). Finally, the dried sample was sputter coated with gold prior to SEM analysis using a scanning electron microscope (Philips XL30) operated at 16 kV. EDX spectra were recorded by focusing on a cluster of the Te nanostructures.

Tellurite reducing sludge collected from the UASB reactor at the end of the experiment was dried at 30°C and made into powder form for Raman spectroscopy and X-ray diffraction (XRD). Micro-Raman spectroscopy (Renishaw inVia Raman microscope, United Kingdom) was performed on the sample to unambiguously distinguish the presence of metallic tellurium. The excitation wavelength of 532 nm from a Nd-YAG laser was used for collecting Raman signals. With this method, a region of interest was identified in the optical image onto which the microscope laser beam was focused for recording the Raman, typically using a 2 µ diameter beam. Powder XRD analysis was carried out with a Bruker D8 Advance using Cu Kα radiation in the range $2\theta = 20 - 90°$.

7.2.4. Extraction and characterization of EPS

Loosely bound-EPS (LB-EPS) was extracted by the centrifugation method (10,000×g for 20 min at 4°C) from control and tellurite-reducing anaerobic granular sludge as described by Bhatia et al. (2013) and D'Abzac et al. (2010). The supernatant was collected as the loosely bound EPS (LB-EPS). The humic and the protein (Lowry modified by Frølund, 1995) as well as the polysaccharide (the sulfuric phenol method by Dubois, 1956) substances in the extracted EPS samples were determined as described earlier by D'Abzac et al. (2010). A

Phoenix 8000 total organic carbon (TOC)-meter (Dohrmann) was used to measure the TOC content of the EPS. The EPS samples were further characterized by recording a three-dimensional excitation-emission matrix (3D-EEM) using a FluoroMax-3 spectrofluorometer (HORIBA Jobin Yvon, Edison, NJ, USA) as described by Bhatia et al. (2013). The EEM spectra were obtained at excitation wavelengths between 240 and 450 nm at 10 nm intervals, and emission wavelengths between 290 and 500 nm.

7.2.5.   Te recovery from tellurite reducing granular sludge

The EPS extraction method was used for recovery of Te associated with the tellurite reducing granular sludge (Yates et al., 2013). The total Te content in the sludge and in the extracted EPS samples were also measured in order to determine the biomass and EPS associated Te(0) as well as the Te(0) recovery efficiency of the centrifugation method. TEM was performed on the extracted EPS samples to determine the morphology of the biogenic Te(0) using a JEOL JEM-100CX II operated at an accelerating voltage of 100 kV. The sludge sample was subsequently subjected to microwave-assisted acid digestion for measuring the total Te concentration.

7.2.6.   Analytical methods

Influent and effluent samples of the UASB reactor were regularly monitored for COD, tellurite, total tellurium and Te(0). COD measurements were carried out using standard methods (APHA 2005). The total Te concentration was analyzed using a graphite furnace atomic absorption spectrophotometer (AAS, SOLAAR MQZe, unity lab services USA) after acidifying samples with concentrated nitric acid (pH < 2). Te(0) was separated from the reactor effluent by adopting the methods as described for elemental Se(0) by Mal et al. (2016b). Te(0) was collected from the effluent sample by centrifuging at 37000 g for 20 min. The pellet was re-suspended in Milli-Q water and Te(0) was measured by using AAS after acidification with concentrated nitric acid (pH < 2). The supernatant was collected for measuring the Te(IV) concentration colorimetrically using the diethyldithiocarbamate (DDTC) method (Turner et al., 1992). Briefly, 0.5 mL sample was added to 1.5 mL 0.5 M TRIS-HCl (pH 7.0) and 0.5 mL 10 mM DDTC reagent. The absorbance was measured at 340 nm using UV-Vis spectrophotometer (UV-2501 PC, Shimadzu).

7.3. Results

7.3.1. Reactor performance – COD removal

The COD removal performance of the UASB reactor as a function of time is shown in Fig. 7.2A. During the start-up period (period I), the COD removal was quickly established and increased from 75 to 91% within the first 7 d of start-up. However, the COD removal efficiency in the UASB reactor dropped to 83% immediately after the addition of 0.08 mM of tellurite. The COD removal improved and reached to > 98% within 2 - 3 weeks of operation with 10 mg. L^{-1} tellurite (period II) (Fig. 7.2A). The COD removal was impacted again when the tellurite concentration was increased to 20 mg. L^{-1} in period III. But, the COD removal efficiency improved within 2 weeks of operation with 0.16 mM tellurite and reached to 93%. The lactate utilization was unaffected by 10 mg. L^{-1} tellurite in period II, while it was marginally impacted by 20 mg. L^{-1} tellurite in period III. However, the lactate utilization increased back to ~98% towards the end of period III.

7.3.2. Reactor performance – removal of Te(IV)

Tellurite removal was noticed from day 1 of addition to the UASB reactor (Fig. 7.2B). But the removal was incomplete at the beginning; therefore, the effluent had a residual tellurite concentration (Fig. 7.2C). The total Te (tellurite and Te(0)) removal efficiency steadily improved within 3 weeks of tellurite addition. Complete and stable removal of tellurite was established towards the end (26 to 48 d) of period II (Fig. 7.2C). The total Te removal efficiency also improved and stabilized at 98% (Fig. 7.2B). The Te that remained in the effluent leaving the reactor was ~0.2 mg Te. L^{-1}, mainly in the form of elemental Te(0) (Fig. 7.2C). Tellurite removal continued in the UASB reactor even when the concentration was increased to 20 mg. L^{-1} TeO$_3^{2-}$. The tellurite and total Te removal efficiency dropped to ~85%, but then steadily increased to 92 and 95%, respectively, towards the end of period III (Fig. 7.2B). The total Te concentration (~1.5 mg. L^{-1}) in the effluent leaving the reactor also increased due to incomplete tellurite reduction (~1.0 mg. L^{-1}) and increase in the biogenic elemental Te(0) concentration (~0.5 mg. L^{-1}) in the effluent (Fig. 7.2C).

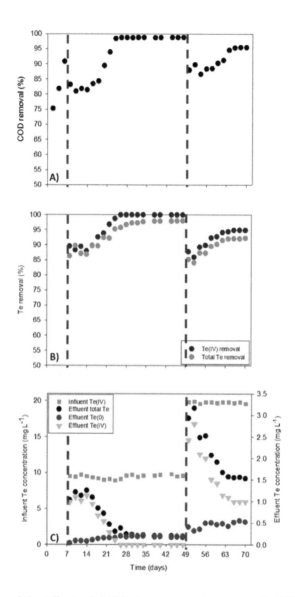

Fig. 7.2 Evolution of the tellurite fed UASB reactor performance. A) COD and B) Tellurite (Te(IV)) and total Te (Te(IV)+Te(0)) removal efficiency. C) Tellurite removal profile and the fate of tellurium in the UASB reactor. The dashed vertical lines divide the three distinct operational periods I, II and III, corresponding to 0, 10 and 20 mg. L^{-1} of Te in form of TeO$_3$$^{2-}$ in the influent, respectively. Legend symbols are inside the corresponding figure

7.3.3.   Characterization of Te associated with granular sludge

Chemical analysis estimated that 49.5 ($\pm$ 3.5) mg Te. g^{-1} DW was retained with the granular sludge in the UASB reactor at the end of the experiment. Low-magnification SEM images revealed the presence of various crystal structures only in the case of the tellurite reducing granular sludge (Fig. 7.3C). EDX confirmed the chemical nature of the crystal deposits to be elemental Te(0) (Fig. 7.3D). The elemental composition (atom %) of the tellurite reducing granules was determined to be C 61.4, O 20.1, Na 2, Te 11.5, P 2.1 and S 2. Signals for elements like carbon, oxygen, nitrogen and sulfur are emanating from granular sludge constituents, i.e. microbial cells, the EPS matrix and other inorganic minerals retained in the granules.

Fig. 7.4A shows the Raman scattering spectrum of the tellurite reducing granular sludge from the UASB reactor. Two characteristic vibration peaks at 121.9 and 140.5 cm^{-1} observed in the Raman spectra of tellurite reducing granular sludge are close to those reported previously for elemental Te(0) (Bonificio & Clarke, 2014; Li et al., 2014). Fig. 7.4B shows a XRD spectrum of tellurite reducing granular sludge which is in good agreement with the standard literature data (COD 1011098) of Te(0) nanocrystals. All the peaks observed in the XRD spectra could be indexed to the presence of hexagonal tellurium in the tellurite reducing granules (Li et al., 2014; Xue et al., 2012).

7.3.4.   Recovery of biogenic Te(0) nanoparticles from UASB granules

In order to investigate the potential to recover Te, the granular sludge was centrifuged to extract LB-EPS and EPS associated Te (Bhatia et al., 2013; D'Abzac et al., 2010; Yates et al., 2013). With the centrifugation (10,000×g for 20 min) method employed in this study, 74-78% of the sludge associated Te was recovered in the supernatant. TEM images of the extracted LB-EPS showed the presence of Te(0) nanostructures: 20-50 nm sized "nanospheres" as well as "nanorods" or clusters of Te(0) "shrads" that coalesced together to form larger 100 – 200 nm clusters (Fig. 7.5).

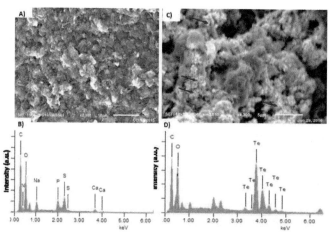

Fig. 7.3 SEM images of control (A) and tellurite reducing (C) anaerobic granular sludge. Arrows indicate Te nanostructures in the tellurite reducing sludge collected from the UASB reactor. EDX spectra of control (B) and tellurite reducing (D) sludge corresponding to the images shown in (A) and (C), respectively

7.3.5. Chemical and EEM fluorescence analysis of Te(0) associated EPS

The EPS extracted from the control and tellurite reducing granules was further characterized by chemical analysis and 3D EEM. The total EPS concentration expressed as TOC DW^{-1} sludge was 21% higher in the tellurite reducing granular sludge than in control sludge. It is evident that the protein (PN) and humic (HS) content was much higher than the polysaccharide (PS) content (Table 7.1). After 70 days of treatment of tellurite containing wastewater, the concentrations of all three EPS components were increased. However, the increase in PN (40%) content was more significant than the HS (16%) and PS (11%) content (Table 7.1).

Two main fluorescence peaks were identified in the 3D-EEM fluorescence spectra of the EPS samples extracted from control and tellurite reducing granules (Fig. 7.6). The first main peak was identified at excitation/emission wavelengths (Ex/Em) of 260-290/310-350 nm (Peak A), while the second main peak was identified at Ex/Em of 410-430/460-480 nm (Peak B). Peak A has been described as a mix of PN-like peaks and soluble microbial products (SMP), in which the fluorescence is mainly associated with the aromatic tyrosine and/or tryptophan protein-like substances and peak B is related to humic-like substances (Romera-Castillo et al.,

163

2011; You et al., 2015; Zhu et al., 2015). Table 7.2 summarizes the fluorescence spectral parameters of the EPS samples, including peak location and fluorescence intensity which was used for quantitative analysis. Peak A appeared to be red-shifted (~10 nm) to longer wavelengths compared to the EPS collected from untreated granular sludge, while there was also a small red shift (~4 nm) in case of Peak B.

Table 7.1. Characterization of EPS extracted from the control and tellurite reducing granular sludge. The extracted EPS samples were analyzed for total protein (PN), polysaccharides (PS) and humic substances (HS).

EPS Samples	TOC (mgC. DW^{-1} sludge)	Biochemical composition of EPS (mg. DW^{-1} sludge)			PN/PS	PN/HS
		PN	PS	HS		
Control sludge	34.08 ± 1.36	31.45 ± 1.95	16.18 ± 1.14	37.66 ± 0.91	1.94	0.84
Te(IV) reducing sludge	41.29 ±1.43	43.99 ± 1.78	18.02 ± 0.93	43.61 ± 1.43	2.42	1.01

7.4. Discussion

7.4.1. Tellurite removal and characterization of biomass associated Te(0)

This study demonstrated for the first time continuous removal of tellurite by anaerobic granules not previously exposed to Te oxyanions and recovery of tellurium from Te-containing synthetic wastewater using UASB reactors. Microorganisms present in the methanogenic granular sludge are capable of Te oxyanion reduction without any prior enrichment (Ramos-Ruiz et al., 2016). A significant improvement in the tellurite removal was nevertheless achieved in the first 3 to 4 weeks of UASB reactor operation, suggesting the need for acclimation of microorganisms towards tellurite reduction or growth of a specific tellurite reducing population (Fig. 7.2). Lactate utilization and COD removal efficiencies dropped immediately after feeding tellurite to the UASB reactor, probably due to the toxicity exerted by tellurite on the microbial community (Zannoni et al., 2008). Fig. 7.2 indicates that lactate utilization and COD removal efficiencies improved along with tellurite removal as microorganisms were able to convert the Te-oxyanion to insoluble elemental form.

Tellurite is highly toxic for most bacteria: the minimum inhibitory concentration (MIC) of tellurite ranges between 0.006 - 0.8 mM for microorganisms such as *Escherichia coli*, *Staphylococcus aureus* and *Pseudomonas aeruginosa* (Harrison et al., 2004; Zannoni et al., 2008). Doubling the tellurite concentration in the influent indeed led to a lower tellurium removal efficiency (~92%) in the UASB reactor (Fig. 7.2). Similar results were reported by Rajwade and Paknikar (2003) for tellurite reduction by *P. mendocina* MCM B-180: the tellurite removal efficiency dropped from 99 to 80% due to the increase in tellurite concentration from 0.08 to 0.2 mM L^{-1} in a continuously stirred tank reactor under aerobic conditions over a 25 day period (Rajwade & Paknikar, 2003). However, tellurite removal by UASB reactors is advantageous in terms of separation of the treated water from the biomass, long term operational stability and no energy requirement for bioreactor aeration.

Te speciation analysis showed that most of the tellurite was removed and only a smaller fraction of Te was present in the treated water leaving the reactor in the form of Te(0) (Fig. 7.2C). The amount of volatile Te in the gas phase was found to be negligible (< 0.5%) compared to the total initial Te concentration (data not shown). These results suggested that the majority of tellurite was reduced to Te(0) and not further to other forms of dissolved or volatile Te, i.e. Te(-II) or organo-Te. Ramos-Ruiz et al. (2016) reported formation of only a minor volatile Te fraction of ~0.14% of the total initial Te concentration in batch experiments of tellurite reduction by anaerobic granular sludge. Similarly the volatile Se fractions were negligible during selenium oxyanion reduction in UASB reactors during the treatment of Se-rich wastewater under comparable operational conditions (Dessì et al., 2016; Lenz et al., 2008).

The negligible presence of volatile Te in the gaseous phase and the presence of Te(0) in low concentrations in the effluent (Fig. 7.2C) suggest that most of the Te(0) formed through tellurite reduction was retained in the granular sludge. Noticeable peaks of Te(0) in the EDX spectrum of the granular sludge confirmed the association of Te(0) on the surface of sludge. The carbon, oxygen, nitrogen and sulfur peaks in the EDX spectrum are likely to be emanating from the EPS and microbial cells in the anaerobic granules and may be attributed to the EPS coating of the biogenic Te(0) nanocrystals (Zonaro et al., 2015). The presence of EPS as capping agent on extracellular Te nanocrystals was also reported in the biogenic production of Te(0) by *Rhodobacter capsulatus* (Borghese et al., 2016). Soda et al. (2011) also reported a similar association of biogenic Se(0) with EPS in the anaerobic granular sludge. Retention of biogenic Se(0) by anaerobic granular sludge during treatment of selenium containing wastewater in UASB reactors has already been suggested as a cost effective selenium recovery process (Dessì et al., 2016; Lenz et al., 2008; Soda et al., 2011). This study shows that the UASB reactor configuration is also suitable for removing tellurium oxyanions and recovering Te(0).

Characterization of the granular sludge by Raman spectroscopy and the presence of distinguished peaks for Te(0) validate the deposition of Te(0) onto the granular sludge (Fig. 7.4A). The presence of the characteristic diffraction peaks of crystalline Te(0) provides further evidence of the formation of Te(0) nanocrystals and their retention in the granular sludge (Fig. 7.4B). However, the lower diffraction intensity of the product indicates that the crystallinity of bioreduced Te(0) nanoparticles was comparatively poor (Li et al., 2014). No other characteristic peaks of impurities such as TeO_2 and K_2TeO_3 were detected in either the Raman spectroscopy or XRD analysis (Fig. 7.4), which also confirms microbial reduction of tellurite to Te(0) by the anaerobic granular sludge.

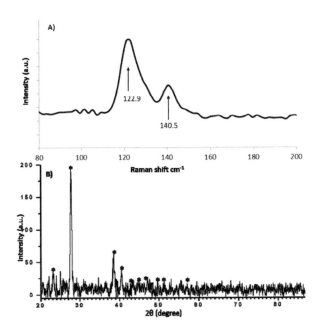

Fig. 7.4 Characterization of Te associated with the tellurite reducing granular sludge harvested from the UASB reactor after 70 d of start-up. Signatures of elemental Te(0) are marked with arrows in the Raman spectra (A) and asterisks in the XRD spectra (B)

7.4.2. Recovery and characterization of biogenic Te(0)

Most of the biomass associated Te was recovered by the centrifugation in the LB-EPS fraction suggesting that biogenic Te(0) was trapped predominantly in the LB-EPS surrounding the sludge. Tellurite reduction and precipitation in the form of biogenic Te(0) occurs in the cytoplasmic space, as well as externally to cells, e.g. on the cell surface and/or periplasmic space (Turner et al., 2012). Depositions of extracellular Te(0) nanoparticles on the cell surface are particularly reported in species such as *Sulfurospirillum barnesii* and *Bacillus beveridgei* using lactate as electron donor (Baesman et al., 2007). Baesman et al. (2009) reported abundant accumulation of Te(0) on the cell surfaces of *B. selenitireducens* as well as the presence of the intracellular Te(0) when grown in the presence of Te(IV). Ramos-Ruiz et al. (2016) also reported the presence of intracellular as well as extracellular Te(0) after Te(IV) reduction by anaerobic granular sludge in batch experiments.

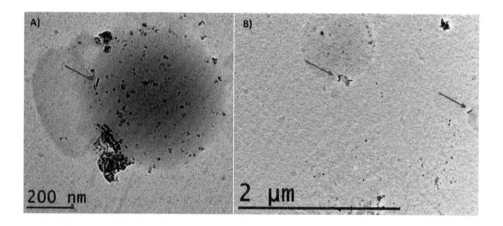

Fig. 7.5 TEM images of the EPS extracted from tellurite reducing anaerobic granular sludge at different magnifications: arrows indicate Te "nano-rod" structures

TEM images of Te(0) nanoparticles in the EPS reveal the presence of Te(0) nanostructures in the form of nanospheres as well as "nanorods" or clusters of Te(0) "shrads" (Fig. 7.5). Formation of clusters of Te shards and individual nanospheres during tellurite reduction by anaerobic granular sludge has been recently reported (Ramos-Ruiz et al., 2016). Baesman et al. (2007) reported the formation of irregularly shaped Te nanospheres with a diameter < 50 nm (also aggregates into larger particles) on the cell surface in the presence of *S. barnesii*.

Precipitation of two different types of Te nanocrystals, i.e. Te-rosettes and Te-granules, occurred simultaneously when a sediment was used for tellurite bioreduction in the presence of lactate (Baesman et al., 2009). Further research on the properties of these biogenic nanomaterials is required to explore the full potential of the biomanufacturing of Te nanoparticles using anaerobic microorganisms.

7.4.3.   Change in EPS composition after treatment of Te-containing wastewater

EPS are composed of complex high-molecular weight mixtures of polymers and have a significant influence on the physicochemical properties of microbial aggregates (Sheng et al., 2010). The EPS matrix of biofilms and microbial aggregates play numerous roles including facilitation of microbial aggregation, entrapment of pollutants, biomineralization and offer protection to microorganisms by acting as physical barrier (Flemming & Wingender, 2010). The major components of EPS are a complex mixture of PN, PS, and HS substances which generally influence the physicochemical properties of sludge in biological wastewater treatment systems (Bhatia et al., 2013; Sheng et al., 2010).

The increase in PN/PS ratio in the EPS extracted from the tellurite reducing UASB sludge suggests a prominent role for PN in tellurite bioreduction, in the protection of granule-constituent microbial cells, as well as in the fate of deposited Te(0) nanoparticles. Wang et al. (2009) have suggested a vital role for proteins in maintaining the structure, properties and functions of microbial aggregates. Cao et al. (2011) have hypothesized a bigger role for the extracellular proteins than polysaccharides in biofilms through electrostatic bonds due to a relatively high content of negatively charged amino acids. Pan et al. (2010) have attributed the role more to proteins as they have a high binding strength to metals.

Moreover, the protein component of the EPS has been reported to be involved in capping and determining the fate (e.g. size and shape) of biogenic Se(0) nanospheres formed by anaerobic granules (Jain et al., 2015; Lenz et al., 2011). It is reported that the elongation factor Tu is one of the most abundant proteins associated to the Se(0) spheres synthesized during selenite reduction by anaerobic granular sludge (Gonzalez-Gil et al., 2016). Elongation factor Tu has its specific amino acid composition and contains 20% more charged residues than the average of other protein families, making it more likely to ionic interactions with the growing Se(0) spheres (Gonzalez-Gil et al., 2016). Hence, further investigation on the interactions of proteins with Te(0) is warranted as the role of proteins in Te(0) biosynthesis and to their fate in the environment cannot be ignored.

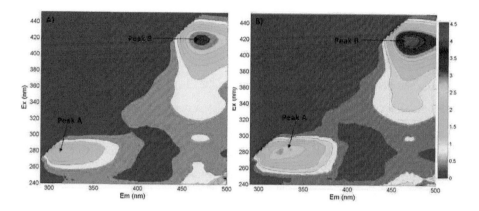

Fig. 7.6 EEM fluorescence spectra of the EPS extracted from control anaerobic granular sludge (A) and tellurite reducing anaerobic granular sludge (B). The tellurite reducing sludge was harvested after 70 d of UASB reactor operation

The EEM spectra of EPS extracted from the anaerobic granular sludge showed a change in the organic structure and components in the EPS due to repeated exposure to tellurite. It is evident that the fluorescence intensity of the aromatic PN-like (tyrosine and/or tryptophan protein-like) and SMP (Pear A) significantly increased (Table 7.2). The importance of aromatic protein-like substances in maintaining the stable structure of the granular sludge has already been reported (Tu et al., 2010; Zhu et al., 2015) and the increase in SMP shows the higher microbial activity in enriched anaerobic granular sludge. Microorganisms produce a higher SMP content under several stress conditions, probably due to the strong metal complexing ability of SMP-like materials (Wang et al., 2015; Wei et al., 2016).

The significant contribution of EPS to both adsorption and reduction has already been reported for metal(loid)s like silver, cadmium and uranium (Li et al., 2016; Raj et al., 2016; Stylo et al., 2013; Wang et al., 2016). The change in fluorescence intensities and shift in peak locations in the EEM fluorescence spectra after the treatment of tellurium containing wastewater indicate that the organic structure and components in EPS were different from the inoculum (Romera-Castillo et al., 2011; Zhu et al., 2015). Although the exact reason behind the changes in EPS composition is unknown, it was suggested that a red shift in Peak A

(Table 7.2) is generally associated with the presence of some functional groups like carbonyl, hydroxyl, amino groups and carboxyl, while any red shift of HS-like (Peak B) fluorescent emission maximum is an indication of increased aromaticity and polycondensation of humic materials (Romera-Castillo et al., 2011; Zhu et al., 2015). These observed shifts in fluorescence peaks could also be due to the formation of Te nanocrystals-EPS complexes (Wang et al., 2016). The results in the present study, including increase in EPS content and particularly the PN-content (Table 7.1) as well as the change in EPS composition (Fig. 7.6) and EPS-associated Te(0) indicate that the EPS play a prominent role in the bioreduction of Te(IV) and synthesis of biogenic Te(0). This study highlights the need for additional studies to understand the changes in composition and fingerprint of EPS upon feeding tellurium oxyanions to UASB granules as well as the EPS-Te interactions to better engineer biosynthesis of Te(0) nanoparticles (Bhatia et al., 2013)

7.4.4. Practical implications

This study shows that reduction of tellurite and entrapment of biogenic Te(0) by the anaerobic granular sludge was the main mechanism of tellurite removal. The biogenic Te(0) was mainly associated with the LB-EPS, and could thus be easily recovered by simple centrifugation, offering possibilities for attractive and cost effective recovery. Further research identifying the microbial population and optimizing the operating conditions (e.g. HRT) are required to improve the reactor performance (Rajwade & Paknikar, 2003). Nevertheless, these findings demonstrate that the UASB reactor technology is an attractive option for the removal and recovery of Te(0) from Te-containing wastewaters. The results obtained in this study can unequivocally help in designing bioreactor operational strategies for microbial reduction of Te oxyanions and recovery of Te(0) at large scale.

7.5. Conclusion

This work demonstrates the continuous tellurite removal and recovery of elemental tellurium using UASB reactors for the first time. More than 98% of the Te was removed from synthetic wastewater containing tellurium (10 mg. L^{-1}). The Te removal efficiencies dropped to 92% when the influent Te concentration increased to 20 mg. L^{-1}. The majority of the Te(VI) was reduced to Te(0) and was associated with the biomass. EDX, Raman spectroscopy and XRD confirm the deposition of biogenic Te(0) on the granular sludge. 78% of the biomass associated Te was recovered by centrifugation in the loosely bound EPS fraction. The EEM fluorescence spectra indicate the organic structure and components in EPS were changed after the Te-containing wastewater treatment, suggesting EPS play a key role in the TeO_4^{2-} removal from the Te-containing wastewater.

Acknowledgements

This research was supported through the Erasmus Mundus Joint Doctorate Environmental Technologies for Contaminated Solids, Soils, and Sediments (ETeCoS^3) (FPA n^0 2010-0009) and the BioMatch project No. 103922 (Role of biofilm-matrix components in the extracellular reduction and recovery of chalcogen) funded by the European Commission Marie Curie International Incoming Fellowship (MC-IIF). The authors would like to thank Dr. Chloé Fourdrin (Université Paris-Est Marne-la-Vallée) for the technical help for Raman spectroscopy and XRD analysis.

References

APHA; AWWA; WPCF; Standard Methods for Water and Wastewater Examination, W., DC, USA 19th ed. 1998.

Ba, L.A., Doring, M., Jamier, V., Jacob, C. 2010. Tellurium: an element with great biological potency and potential. *Org Biomol Chem.*, **8**, 4203–4216.

Baesman, S.A., Bullen, T.D., Dewald, J., Zhang, D., Curran, S., Islam, F.S., Beveridge, T.J., Oremland, R.S. 2007. Formation of tellurium nanocrystals during anaerobic growth of bacteria that use te oxyanions as respiratory electron acceptors. *Appl Environ Microbiol.*, **73**(7), 2135-2143.

Baesman, S.M., Stolz, J.F., Kulp, T.R., Oremland, R. 2009. Enrichment and isolation of *Bacillus beveridgei* sp. nov., a facultative anaerobic haloalkaliphile from Mono Lake, California, that respires oxyanions of tellurium, selenium, and arsenic. *Extremophiles.*, **13**(4), 695-705.

Belzile, N., Chen, Y.-W. 2015. Tellurium in the environment: A critical review focused on natural waters, soils, sediments and airborne particles. *Appl Geochem.*, **63**, 83-92.

Bhatia, D., Bourven, I., Simon, S., Bordasa, F., E.D., v.H., Rossano, S., Lens, P.N.L., Guibaud, G. 2013. Fluorescence detection to determine proteins and humic-like substances fingerprints of exopolymeric substances (EPS) from biological sludges performed by size exclusion chromatography (SEC). *Bioresour Technol.*, **131**, 159-165.

Bonificio, W.D., Clarke, D.R. 2014. Bacterial recovery and recycling of tellurium from tellurium-containing compounds by *Pseudoalteromonas* sp. EPR3. *J Appl Microbiol.*, **117**(5), 1293-1304.

Borghese, R., Brucale, M., Fortunato, G., Lanzi, M., Mezzi, A., Valle, F., Cavallini, M., Zannoni, D. 2016. Extracellular production of tellurium nanoparticles by the photosynthetic bacterium Rhodobacter capsulatus. *J Hazard Mater.*, **309**, 202-209.

Cao, B., Shi, L., Brown, R.N., Xiong, Y., Fredrickson, J.K., Romine, M.F., Marshall, M.J., Lipton, M.S., Beyenal, H. 2011. Extracellular polymeric substances from Shewanella sp. HRCR-1 biofilms: characterization by infrared spectroscopy and proteomics. *Environ Microbiol.*, **13**(4), 1018-1031.

Cunha, R.L., Gouvea, I.E., Juliano, L. 2009. A glimpse on biological activities of tellurium compounds. *An Acad Bras Cienc.*, **81**(3), 393-407.

D'Abzac, P., Bordas, F., van Hullebusch, E.D., Lens, P.N.L., Guibaud, G. 2010. Extraction of extracellular polymeric substances (EPS) from anaerobic granular sludges: comparison of chemical and physical extraction protocols. *Appl Microbiol Biotechnol.*, **85**, 1589-1599.

de Boer, M.A., Lammertsma, K. 2013. Scarcity of rareearth elements. *Chemsuschem.*, **6**(11), 2045-2055.

Deng, Z.T., Zhang, Y., Yue, J.C., Tang, F.Q., Wei, Q. 2007. Green and orange CdTe quantum dots as effective pH-sensitive fluorescent probes for dual simultaneous and independent detection of viruses. *J Phys Chem. B*, **111**(41), 12024-12031.

Dessì, P., Jain, R., Singh, S., Seder-Colomina, M., van Hullebusch, E.D., Rene, E.R., Ahammad, S.Z., Lens, P.N.L. 2016. Effect of temperature on selenium removal from wastewater by UASB reactors. *Water Res.*, **94**, 146-154.

Flemming, H., Wingender, J. 2010. The biofilm matrix. *Nat Rev Microbiol.*, **8**, 623-633.

Forootanfar, H., Amirpour-Rostami, S., Jafari, M., Forootanfar, A., Yousefizadeh, Z., Shakibaie, M. 2015. Microbial-assisted synthesis and evaluation the cytotoxic effect of tellurium nanorods. *Mater Sci Eng C Mater Biol Appl.*, **49**, 183-189.

Gonzalez-Gil, G., Lens, P.N.L., Saikaly, P.E. 2016. Selenite reduction by anaerobic microbial aggregates: Microbial community structure, and proteins associated to the produced selenium spheres. *Front Microbiol.*, **7**, 571.

Gonzalez-Gil, G., Seghezzo, L., Lettinga, G., Kleerebezem, R. 2001. Kinetics and mass-transfer phenomena in anaerobic granular sludge. *Biotechnol Bioeng.*, **73**(2), 125-134.

Harrison, J.J., Ceri, H., Stremick, C.A., Turner, R.J. 2004. Biofilm susceptibility to metal toxicity. *Environ Microbiol.*, **6**, 1220-1227.

Jain, R., Jordan, N., Schild, D., van Hullebusch, E.D., Weiss, S., Franzen, C., Farges, F., Hübner, R., Lens, P.N.L. 2015. Extracellular polymeric substances govern the surface charge of biogenic elemental selenium nanoparticles. *Environ Sci Technol.*, **49**, 1713-1720.

Kim, D.H., Kanaly, R.A., Hur, H.G. 2012. Biological accumulation of tellurium nanorod structures via reduction of tellurite by *Shewanella oneidensis* MR-1. *Bioresour Technol.*, **125**, 127-131.

Klonowska, A., Heulin, T., Vermeglio, A. 2005. Selenite and tellurite reduction by Shewanella oneidensis. *Appl Environ Microbiol.*, **71**, 5607-5609.

Lenz, M., E.D., V.H., Hommes, G., Corvini, P.F., Lens, P.N.L. 2008. Selenate removal in methanogenic and sulfate-reducing upflow anaerobic sludge bed reactors. *Water Res.*, **42**, 2184–2194.

Lenz, M., Kolvenbach, B., Gygax, B., Moes, S., Corvini, P.F. 2011. Shedding Light on Selenium Biomineralization: Proteins Associated with Bionanominerals. *Appl Environ Microbiol.*, **77**(13), 4676-4680.

Li, G., Cui, X., Tan, C., Lin, N. 2014. Solvothermal synthesis of polycrystalline tellurium nanoplates and their conversion into single crystalline nanorods. *RSC Adv.*, **4**, 954-960.

Li, P., Liu, S., Yan, S., Fan, X., He, Y. 2011. A sensitive sensor for anthraquinone anticancer drugs and hsDNA based on CdTe/CdS quantum dots fluorescence reversible control. *Colloids Surf. A.*, **392**, 7-15.

Li, S.W., Zhang, X., Sheng, G.P. 2016. Silver nanoparticles formation by extracellular polymeric substances (EPS) from electroactive bacter. *Environ Sci Pollut Res Int.*, **23**(9), 8627-8633.

Mal, J., Nancharaiah, Y.V., van Hullebusch, E.D., Lens, P.N.L. 2016. Effect of heavy metal co-contaminants on selenite bioreduction by anaerobic granular sludge. *Bioresour Technol.*, **6**, 1-8.

Nancharaiah, Y.V., Mohan, S.V., Lens, P.N.L. 2016. Biological and bioelectrochemical recovery of critical and scarce Metals. *Trends Biotechnol.*, **34**(2), 137-155.

Ngwenya, N., Whiteley, C.G. 2006. Recovery of rhodium (III) from solutions and industrial wastewaters by a sulfate-reducing bacteria consortium. *Biotechnol Prog.*, **22**, 1604-1611.

Pan, X., Liu, J., Zhang, D., Chen, X., Song, W., Wu, F. 2010. Binding of dicamba to soluble and bound extracellular polymeric substances (EPS) from aerobic activated sludge: a fluorescence quenching study. *J Colloid Interface Sci.*, **345**, 442-447.

Raj, R., Dalei, K., Chakraborty, J., Das, S. 2016. Extracellular polymeric substances of a marine bacterium mediated synthesis of CdS nanoparticles for removal of cadmium from aqueous solution. *J Colloid Interface Sci.*, **462**, 166-175.

Rajwade, J.M., Paknikar, K.M. 2003. Bioreduction of tellurite to elemental tellurium by *Pseudomonas mendocina* MCM B-180 and its practical application. *Hydrometallurgy*, **71**(1), 243-248.

Ramos-Ruiz, A., Field, J.A., Wilkening, J.V., Sierra-Alvarez, R. 2016. Recovery of elemental tellurium nanoparticles by the reduction of tellurium oxyanions in a methanogenic microbial consortium. *Environ Sci Technol.*, **50**(3), 1492-500.

Romera-Castillo, C., Sarmento, H., A´lvarez-Salgado, X.A., Gasol, J.M., Marrase, C. 2011. Net Production and Consumption of Fluorescent Colored Dissolved Organic Matter by Natural Bacterial Assemblages Growing on Marine Phytoplankton Exudates. *Appl Environ Microbiol.*, **77**(21), 7490-7498.

Sheng, G., Yu, H.Q., Li, X.Y. 2010. Extracellular polymeric substances (EPS) of microbial aggregates in biological wastewater treatment systems: a review. *Biotechnol Adv.*, **28**, 882-894.

Soda, S., Kashiwa, M., Kagami, T., Kuroda, M., Yamashitab, M., Ike, M. 2011. Laboratory-scale bioreactors for soluble selenium removal from selenium refinery wastewater using anaerobic sludge. *Desalination*, **279**, 433-438

Sredni, B., Geffen-Aricha, R., Duan, W., Albeck, M., Sonino, T., Longo, D.L., Mattson, M.P., Yadid, G. 2007. Multifunctional tellurium molecule protects and restores dopaminergic neurons in Parkinson's disease models. *FASEB J.*, **21**(8), 1870-1883.

Stams, A.J.M., Grolle, K.C.F., Frijters, C.T.M.J., van Lier, J.B. 1992. Enrichment of thermophilic propionate-oxidizing bacteria in syntrophy with Methanobacterium thermoautotrophicum or Methanobacterium thermoformicicum. *Appl Environ Microbiol.*, **58**, 346-352.

Stylo, M., Alessi, D.S., Shao, P.P., Lezama-Pacheco, J.S., Bargar, J.R., Bernier-Latmani, R. 2013. Biogeochemical controls on the product of microbial U(VI) reduction. *Environ Sci Technol.*, **47**, 12351-12358.

Tanaka, M., Arakaki, A., Staniland, S.S., Matsunaga, T. 2010. Simultaneously discrete biomineralization of magnetite and tellurium nanocrystals in Magnetotactic Bacteria. *Appl Environ Microbiol.*, **76**(16), 5526-5532.

Tu, X.A., Su, B.S., Li, X.N., Zhu, J.R. 2010. Characteristics of extracellular fluorescent substances of aerobic granular sludge in pilot-scale sequencing batch reactor. *J Central South Univ Technol.*, **17**(3), 522-528.

Turner, R.J., Borghese, R., Zannoni, D. 2012. Microbial processing of tellurium as a tool in biotechnology. *Biotechnol Adv.*, **30**(5), 954-963.

Turner, R.J., Weiner, J.H., Taylor, D.E. 1992. Use of diethyldithiocarbamate for quantitative determination of tellurite uptake in bacteria. *Anal Biochem.*, **2014**, 292-295.

Wang, Q., Kang, F., Gao, Y., Mao, X., Hu, X. 2016. Sequestration of nanoparticles by an EPS matrix reduces the particlespecific bactericidal activity. *Sci Rep.*, **9**, 21379-21389.

Wang, Y., Qin, J., Zhou, S., Lin, X., Ye, L., Song, C., Yan, Y. 2015. Identification of the function of extracellular polymeric substances (EPS) in denitrifying phosphorus removal sludge in the presence of copper ion. *Water Res.*, **73**, 252-264.

Wang, Z., Wu, Z., Tang, S. 2009. Characterization of dissolved organic matter in a submerged membrane bioreactor by using three-dimensional excitation and emission matrix fluorescence spectroscopy. *Water Res.*, **43**(6), 1533-1540.

Wei, D., Dong, H., Wu, N., Ngo, H.H., Guo, W., Du, B., Wei, Q. 2016. A fluorescence approach to assess the production of soluble microbial products from aerobic granular sludge under the stress of 2,4-Dichlorophenol. *Sci Rep.*, **6**, 24444.

Xue, F., Bi, N., Liang, J., Han, H. 2012. A simple and efficient method for synthesizing Te Nanowires from CdTe nanoparticles with EDTA as shape controller under hydrothermal condition. *J Nanomater.*, **2012**. doi.org/10.1155/2012/751519

Yates, M.D., Cusick, R.D., Logan, B.E. 2013. Extracellular Palladium Nanoparticle Production using *Geobacter sulfurreducens*. *ACS Sustainable Chem Eng.*, **1**, 1165-1171.

You, G., Hou, J., Xu, Y., Wang, C., Wang, P., Miao, L., Ao, Y., Li, Y., Lv, B. 2015. Effects of CeO_2 nanoparticles on production and physicochemical characteristics of extracellular polymeric substances in biofilms in sequencing batch biofilm reactor. *Bioresour Technol.*, **194**, 91-98.

Zannoni, D., Borsetti, F., Harrison, J., Turner, R. 2008. The bacterial response to the chalcogen metalloids Se and Te. . *Adv Microb Physiol.*, **53**, 1-12.

Zhu, L., Zhou, J., Lv, M., Yu, H., Zhao, H., Xu, X. 2015. Specific component comparison of extracellular polymeric substances (EPS) in flocs and granular sludge using EEM and SDS-PAGE. *Chemosphere.*, **121**, 26-32.

Zonaro, E., Lampis, S., Turner, R.J., Qazi, S.J., Vallini, G. 2015. Biogenic selenium and tellurium nanoparticles synthesized by environmental microbial isolates efficaciously inhibit bacterial planktonic cultures and biofilms. *Front Microbiol.*, **6**, 584-595.

CHAPTER 8

A comparison of fate and toxicity of selenite, biogenically and chemically synthesized selenium nanoparticles to the Zebrafish (*Danio rerio*) embryogenesis

This chapter has been published in modified form:

Mal, J., Veneman, W., Nancharaiah, Y.V., van Hullebusch, E.D., Peijnenburg, W., Vijver, M.G., Lens, P.N.L. 2017. A comparison of fate and toxicity of selenite, biogenically and chemically synthesized selenium nanoparticles to the Zebrafish (*Danio rerio*) embryogenesis. *Nanotoxicology*. 11, 1, 87-97.

Abstract

Microbial reduction of Se oxyanions to elemental Se is a promising technology for bioremediation and treatment of Se wastewaters. But, a fraction of biogenic nano-Selenium (nano-Se[b]) formed in bioreactors remain suspended in the treated waters, thus entering the aquatic environments. The present study investigated the toxicity of nano-Se[b] formed by anaerobic granular sludge biofilms on zebrafish embryos in comparison with selenite and chemogenic nano-Se (nano-Se[c]). The nano-Se[b] formed by granular sludge biofilms showed a LC50 value of 1.77 mg. L^{-1} was 3.2-fold less toxic to zebrafish embryos than selenite (LC50=0.55 mg. L^{-1}) and 10-fold less toxic than bovine serum albumin stabilized nano-Se[c] (LC50=0.16 mg. L^{-1}). Moreover, smaller (nano-Se[cs]; particle diameter range: 25-80 nm) and larger (nano-Se[cl]; particle diameter range: 50-250 nm) sized nano-Se[c] particles showed comparable toxicity on zebrafish embryos. The lower toxicity of nano-Se[b] in comparison to nano-Se[c] was analyzed in terms of the stabilizing organic layer. The results confirmed that the organic layer extracted from the nano-Se[b] consisted of components of the extracellular polymeric substances (EPS) matrix, which govern the physiochemical stability and surface properties like ζ-potential of nano-Se[b]. Based on the data, it is contented that the presence of humic acid like substances of EPS on the surface of nano-Se[b] plays a major role in lowering the bioavailability (uptake) and toxicity of nano-Se[b] by decreasing the interactions between nanoparticles and embryos.

Key words: Biogenic nanomaterials, selenium nanoparticles, zebrafish, EPS, nanotoxicology

8.1. Introduction

Selenium (Se) is an essential micronutrient in organisms including humans. It plays a very important role in several cellular processes including thyroid hormone production and in mitigating oxidative stress (Rayman, 2006). However, at concentrations of one order excess of its essential level, Se becomes a potential toxic trace element (Lenz & Lens, 2009; Winkel et. al., 2012). Se contamination of aquatic bodies generally occurs through discharge of different waste streams such as acid mine drainage, coal and phosphate mining effluents, flue gas desulphurization waters and agriculture drainage (Mal et. al., 2016; Nancharaiah & Lens, 2015a). Elemental Se (Se(0)) is an integral component of the natural Se cycle often formed in oxygen limited environments, with microorganisms playing a key role in bioreduction of Se oxyanions (selenite and selenate) (Winkel et. al, 2012). In fact, conversion of soluble Se oxyanions to insoluble Se(0) is a key process for *in situ* bioremediation and treatment of Se contaminated waters and wastewaters (Nancharaiah & Lens, 2015a; 2015b). But a significant fraction of Se(0) formed by microbial reduction remains in the bioreactor effluents leading to the discharge of Se(0) in the environment. Abiotic reduction of Se-oxyanions under reducing environments, by e.g. Fe(II)-containing mineral surfaces may also lead to significant deposition of Se(0) in natural environments as sediments, groundwater or nuclear waste disposal geological media (Scheinost et. al., 2008).

Se pollution of aquatic environments is a concern mainly because this element can cross trophic levels of the food-chain and gets biomagnified, showing an unusual high bioaccumulation potential and exhibiting teratogenic effects on aquatic organisms (Dennis, 2004). Moreover, the toxicity of Se on aquatic organisms depends on its chemical (inorganic and organic) form, physical (liquid, solid and gas) form and concentration. Biogenic elemental Se (nano-Se[b]) is always in the form of nanoparticles (nano-Se) with a maximum size up to ~400 nm, thus bringing nanotoxicology aspects (e.g. dissolution, entry into cells, surface chemistry and reactivity) into consideration (Winkel et al., 2012; Buchs et al., 2013). Although the average particle size of nano-Se[b] is more than 100 nm, microbially produced Se(0) is often referred to as nano-Se in the literature (Jain et al., 2015; Srivastava et al., 2013; Zhang et al., 2011). In addition, the particles of nano-Se[b] formed by microorganisms are colloidal in nature (Buchs et al., 2013); hence the term nano-selenium is used in this paper.

Compared to soluble oxyanions, Se(0) is considered to be less bioavailable and reactive. However, some researchers have reported that the uptake of Se(0) by marine bivalve mollusks such as *Macoma balthica* or *Potamocorbula amurensis* (Luoma et al., 2000; Schlekat et al., 2000). *M. balthica* ingests sediments with >1.5 µg selenium g^{-1}, a concentration that could achieve steady-state tissue burdens approaching the level toxic to fish. Thus, it is necessary that treatment of selenium containing waters by stimulation of microbial dissimilatory reduction of selenium oxyanions should be considered in efforts to avoid discharge and to reduce hazardous exposures to selenium at highly contaminated sites. However, studies on the fate and toxicity of nano-Se, particularly on nano-Se[b] in environmentally relevant conditions are limited and the available information is contradictory. Chemically synthesized nano-Se (nano-Se[c]) was reported to be 7-fold less toxic than selenite when exposed orally to mice and rats (Zhang et al., 2001). But, the bioavailability (uptake) of nano-Se[c] was comparable to that of selenite. A subsequent study showed a 5-fold higher toxicity for nano-Se[c] than selenite on juvenile medaka fish due to hyper-accumulation and slow clearance of nano-Se[c] from the liver (Li et al., 2008).

The colloidal stability of nano-Se[b] is governed by the associated organic layer, which originates from the microorganisms and their extracellular polymeric substances (EPS) (Buchs et al., 2013; Jain et. al., 2015). In natural and engineered settings, the EPS matrix of biofilms plays a critical role in determining the transport, fate and toxicity of engineered nanoparticles (NPs). It has been suggested that the natural organic matter (NOM) available in surface and ground waters influences the bioavailability of selenium oxyanion (e.g. selenite) (Zhang & Moore, 1996). But, it is unclear whether NOM or EPS of biofilms associate with elemental selenium and influence the nano-Se bioavailability and toxicity.

In the present study, experimental work was designed to investigate the toxicity of nano-Se[b] formed in bioreactors treating selenium wastewater. Zebrafish embryos were chosen because zebrafish is an ideal vertebrate model organism for studying the effects of environmental contaminants on developmental processes. In addition, the transparency of the zebrafish embryos allows continuous observation of developmental changes during organogenesis (Lohr & Yost, 2000). The toxicity of nano-Se[b] was compared with those of nano-Se[c] and selenite. Two different sizes of nano-Se[c] were taken into consideration to determine the effect of particle size on toxicity. The fate of nano-Se was determined in terms of shape, size, aggregation and dissolution kinetics. The organic constituents of the surface coating associated with nano-Se[b] was characterized in order to better understand the importance of the surface coating and compared its influence on the fate and toxicity of nano-Se to the zebrafish embryos with bovine serum albumin (BSA)-capped nano-Se[c].

8.2. Materials and methods

8.2.1. Chemicals

Sodium selenite (Na_2SeO_3) was purchased from Sigma-Aldrich and was used as received. Nano-Se[b] and nano-Se[c] were prepared by biological and chemical reduction, respectively, as described below.

8.2.2. Nano-Se production

Nano-Se[b] was produced through bioreduction of selenite by anaerobic granular sludge. Anaerobic granular sludge collected from a full scale upflow anaerobic sludge blanket (UASB) reactor treating paper mill wastewater (Industriewater Eerbeek B.V., Eerbeek, The Netherlands) was utilized as the source of microorganisms for selenite reduction. The mineral medium used for biogenic selenium synthesis contained (mg. L[-1]): NH_4Cl (300), $CaCl_2.2H_2O$ (15), KH_2PO_4 (250), Na_2HPO_4 (250), $MgCl_2$ (120), and KCl (250). Sodium lactate (5 mM) and Na_2SeO_3 (1 mM) were used as the electron donor and electron acceptor, respectively. The pH of the mineral medium was adjusted to 7.3. The serum bottles with 400 mL mineral medium and granular sludge (4.0 g wet weight) were closed with butyl rubber septa, then purged with N_2 gas for ~5 min and incubated at 30°C for 7 days in an orbital shaker set at 150 rpm. Formation of nano-Se[b] was confirmed by the appearance of a red color and disappearance of selenite. After the incubation period, the supernatant was decanted from the serum bottles and the suspended biomass was removed by centrifugation (Hermle Z36K) at 3000 g, 4 °C for 15 min. The supernatant was centrifuged again at 37,000 g, 4 °C for 15 min. The red colored pellet was suspended in Milli-Q (18MΩ cm) water by sonication and purified by hexane (Jain et al., 2015).

Nano-Se[c] was produced through chemical reduction of selenite as described by Zhou et al. (2014). Briefly, selenite (0.1 M, 500 μL) was mixed with a 2.5 mL solution containing 15.5 mg of glutathione and 10 mg of BSA for production of smaller sized nano-Se (nano-Se[cs]) particles. Larger sized nano-Se (nano-Se^{cl}) particles were prepared by using 15.5 mg of glutathione and 4 mg of BSA. Within five minutes, the color of the solution turned from colorless to red due to the formation of nano-Se^c.

The nano-Se^b and nano-Se^c (nano-Se^{cs} and nano-Se^{cs}) particles were collected by centrifugation at 37,000 g, 4°C for 30 min. The pellets were dried at 30°C to obtain nano-Se in the powder form. Heating was avoided as amorphous Se(0) undergoes a glass transition at temperatures above 30-35°C (Pearce et al., 2009).

8.2.3.   Stock solutions of selenite, nano-Se^b, and nano-Se^c

The selenite stock was prepared by dissolving 1 gL-1 of sodium selenite (Na$_2$SeO$_3$, 99% purity) in ultrapure water. Nano-Se^b, nano-Se^{cs} and nano-Se^{cl} stocks were prepared by suspending, respective, 100 mg. L^{-1} powder in ultrapure water. The suspensions were sonicated for 15 min using S40 h Elmasonic water bath sonicator. The Se concentration was determined in the stock solutions by inductively coupled plasma-mass spectrometry (ICP-MS) after acidification in 0.6 M HNO$_3$.

8.2.4.   Characterization of nano-Se

The size and morphology of the nano-Se^b, nano-Se^{cs} and nano-Se^{cl} suspension in egg water (0.21 g Instant Ocean® salt in 1 L Milli-Q water, pH 6.5-7.0) after 1 and 24 h of incubation were characterized using transmission electron microscope (TEM) (JEOL 1010). For sample preparation, a drop of egg water containing nano-Se was deposited onto a 400 mesh Cu grid coated with a carbon support film. After drying at room temperature, TEM analysis was performed. Dynamic light scattering of samples was performed on a Zetasizer Nano-ZS instrument (Malvern Instruments) for determining the size distribution and zeta (ζ)-potential of nano-Se particles suspended in egg water. Nano-Se suspensions having 1 and 5 mg. L^{-1} Se were prepared in egg water. Samples were drawn at 1h and 24 h of preparation for measurements. A maximum incubation of 24 h was chosen because the medium of the Zebra embryos assay was changed every 24 h. Nano-Se^b was also characterized by scanning electron microscope (SEM) equipped with an energy dispersive X-ray spectra system (EDXS) (see Appendix 1 for more details).

8.2.5.   Characterization of organic material associated with nano-Se

Organic material was extracted from nano-Se[b] and nano-Se[c] using the formaldehyde (0.06 mL of 36.5% formaldehyde, at 4°C for 1 h) plus NaOH (1 N; at 4°C for 3 h) extraction method (Liu & Fang, 2002). The extracted sample was centrifuged at 20,000 g for 20 min, then filtered with a 0.22 μm acetate filter and finally diluted in water to bring the dissolved total organic carbon concentration to ~10 mg. L^{-1}. Organic material in the diluted extract was characterized by three-dimensional excitation-emission matrix (3D-EEM) using a FluoroMax-3 spectrofluorometer (HORIBA Jobin Yvon, Edison, NJ, USA) (Maeng et. al., 2012; Bhatia et. al., 2013). The EEM spectra were obtained at excitation wavelengths between 240 and 450 nm at 10 nm intervals, and emission wavelengths between 290 and 500 nm. The 3D spectra were divided and expressed into five regions, each region associated with different compounds: derived from proteins - tyrosine (I) and tryptophan (II) corresponding to aromatic proteins, fulvic-like acids (III), soluble microbial by-products (SMP) (IV) and humic-like substances (V) (Maeng et al., 2012; Leenheer & Croué, 2003; Baker & Lamont-Black, 2001).

8.2.6.   Dissolution kinetics of nano-Se

The dissolution of nano-Se, if any, was determined in exposure medium (i.e. egg water) used for toxicity tests. Suspension of nano-Se[b] and nano-Se[c] (nano-Se[cs] and nano-Se[cl], both) at 1 mg. L^{-1} Se were prepared in exposure medium and incubated for a maximum period of 24 h. At different time intervals (0, 8 and 24 h), the dissolved Se concentration was determined. The suspension containing nano-Se[b] and nano-Se[c] were first centrifuged at 40000 g for 30 min at 4°C (Hermle Z36K). Supernatants were filtered through 0.1 μm syringe filter (Antop 25, Whatman). Se concentrations were measured in the filtrates using atomic absorption spectrophotometer (AAS) (SOLAAR MQZE, Unity lab services USA) after acidification in 0.6 M HNO$_3$. The presence of selenite and selenate in supernatants was monitored at 24 h to confirm any reoxidation of nano-Se[b] or nano-Se[c] to selenite or selenate. Selenate was determined by Ion Chromatography (IC), equipped with an AS4A column with the retention time of selenate at 10.3 min. For selenite analysis, a spectrophotometric method was followed (Mal et al., 2016). Briefly, the supernatant (1 mL) was mixed with 0.5 mL of 4 M HCl, and then with 1 mL of 1 M ascorbic acid. After 10 min of incubation at room temperature, the absorbance was determined at 500 nm using an UV-Vis spectrophotometer (Hermle Z36 HK).

Chapter 8

8.2.7. Toxicity assay of selenite, nano-Se[b] and nano-Se[c] on zebrafish embryos

ABxTL wild-type adult zebrafish was maintained at $25 \pm 5°C$ in a 14 h light: 10 h dark cycle. Fertilized zebrafish eggs obtained from adults were distributed into 96-well plates at 1 egg each per well. An acute exposure regime of 96 h was used, from 24 to 120 h post fertilization (hpf), thus including the major stages of organ development. At 24 hpf, the embryos were exposed to 250 µL/well of freshly prepared egg water containing Na_2SeO_3, nano-Se[b], nano-Se[cs] and nano-Se[cl]. Nominal concentrations of 0.2, 0.4, 0.6, 0.8 and 1.0 mg. L^{-1} of Na_2SeO_3 and 0.05, 0.1, 0.2, 0.5, 1, 2 and 5 mg. L^{-1} of nano-Se[b], nano-Se[cl] or nano-Se[cs] were prepared in egg water. The suspensions were sonicated for 15 min in an S40 h Elmasonic water bath sonicator and then exposed to embryos immediately. Multi-well plates containing embryos in egg water were kept at 28°C. The exposure medium (200 µL) was always replaced with a freshly prepared medium containing either selenite or nano-Se every day up to 120 hpf according to Organization for Economic Co-operation and Development (OECD) guideline 157 (OECD, 2011).

Duplicate trials containing 8 embryos per treatment group were used. Mortality and hatching rate were determined at 48, 72, 96 and 120 hpf using a dissecting stereomicroscope (Leica M165C). Embryos were scored as dead based on criteria according to OECD guideline 157. For the median lethal concentration (LC_{50}) calculation, mortality data obtained at 120 hpf was used. The LC_{50} was determined based on plotting mortality data on a cumulative curve obtained from two independent experiments at 120 hpf. The LC_{50} values of zebrafish embryos for selenite and nano-Se were calculated using a non-linear dose response function available in GraphPad Prism 5:

$$E = \frac{Bottom+(Top-Bottom)}{1+\frac{10(logLC50-logC)}{\rho}}$$ (8.1)

where E is the mortality effect on zebrafish embryos (scaled 0-1) and top and bottom are plateaus in the units of the y axis, with the top and bottom of mortality being 100% and 0%, respectively. The term C is the initial actual exposure concentration of selenite or nano-Se, and ρ is the slope of the curve.

The toxicity of nano-Se[b] and nano-Se[c] to zebrafish embryos was determined as total nano-Se[b]$_{total}$ and nano-Se[c]$_{total}$ which is the sum of the contributions of both nano-Se$_{particle}$ and nano-

184

Se_{ion}. The actual concentrations of nano-Se_{ion} and nano-Se_{total} were measured by AAS at the end of 24 h incubation in the exposure medium as described previously. The toxicity of nano-Se_{ion} can be measured by the concentration-response curve of Na_2SeO_3. Therefore, mortality of embryos induced solely by nano-$Se_{particle}$ was calculated by using the response addition model (Backhaus et. al., 2000):

$$E_{total} = 1 - [(1 - E_{ion})(1 - E_{particles})] \qquad (8.2)$$

where E_{total}, E_{ion} and $E_{particle}$ represent the effect on zebrafish embryos caused by the nano-Se_{total}, nano-Se_{ion} and nano-$Se_{particle}$ (scaled 0–1), respectively which makes $E_{particle}$ as the only unknown, allowing for direct calculation of the effect caused by the particles at any specific initial actual particle concentration.

8.2.8. Statistical analysis

All the data was expressed as mean with a 95% confidence interval (C.I.) or standard error of the mean. To compare the differences between treatments and controls, statistical analysis using one-way analysis of variance (ANOVA) with Tukey's multiple comparison posttests was performed when required. The significance level in all the calculations was set at $p < 0.05$.

8.3. Results

8.3.1. Characterization of nano-Se

Nano-Se^b produced during the bioreduction of selenite by anaerobic granular sludge was spherical in shape as shown by TEM (Fig. 8.1) and SEM (Fig. S8.1, Appendix 1) images. EDXS analysis confirmed the presence of selenium as the main element in nano-Se^b samples (Fig. S8.1, Appendix 1). In addition, the presence of peaks for carbon, nitrogen, and oxygen were attributed to the organic material associated with the nano-Se^b. The morphology of the nano-Se^c was also spherical as revealed using TEM imaging (Fig. 8.1). The particle size distribution and zeta-potential of all three kinds of nano-Se suspensions in egg water determined after 1 h and 24 h of preparation are given in Table 8.1. Nano-Se^b particles had sizes ranging from 100 to 350 nm with an average size of 185 nm (Fig. S8.2, Appendix 1). The size of nano-Se^{cs} particles, formed in the presence of 10 mg. L^{-1} of BSA, ranged from 25 to 90 nm with an average diameter of 50 nm, while nano-Se^{cl} formed at lower BSA concentrations (4 mg L^{-1}) was larger in size and ranged from 50 to 250 with an average diameter of 106 nm (Fig. S8.2, Appendix 1). It was apparent from the data that no significant change in particle size distribution was seen when the particles were suspended in egg water even after 24 h (Fig. S8.1, Appendix 1). The zeta potential of the nano-Se^b (-28.5 ± 5.19mV) suspensions was higher than both of nano-Se^{cs} (-16.2 ± 3.16mV) and nano-Se^{cl} (-13.73 ± 3.57mV). There was no marked change in the zeta potential of nano-Se suspended in egg water during 24 h incubation (Table 8.1).

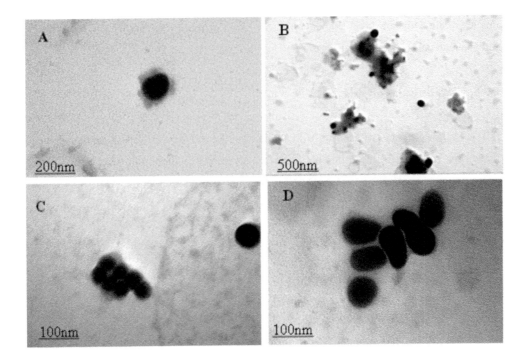

Fig. 8.1 Transmission electron microscopic images of nano-Se particles in egg water; A-B) nano-Se^b, C) nano-Se^{cs} and D) nano-Se^{cl}

8.3.2. Dissolution kinetics of nano-Se

Release of Se ions, if any, from nano-Se in egg water was determined immediately, and after 8 and 24 h of incubation (Fig. S8.3, Appendix 1). Dissolution of nano-Se^{cs} (5.6±0.8%) at 0 h was relatively higher than nano-Se^{cl} (4.4±0.17%) and nano-Se^b (1.3±0.1%). But, dissolution at

24 h of nano-Se^{cl} was comparatively higher than both nano-Se^{cs} and nano-Se^b. The dissolved Se concentration after 24 h was 3.07(±0.12), 7.76(±0.26) and 8.4(±0.4)% of the initial Se (1 mg. L^{-1}), respectively, for nano-Se^b, nano-Se^{cs}, and nano-Se^{cl}.

8.3.3.   Fluorescence properties of organic material associated with nano-Se

The 3D EEM fluorescence spectra corresponding to different components of organic material extracted from nano-Se^c (nano-Se^{cs} and nano-Se^{cl}, both have only BSA as capping agent) and nano-Se^b are shown in Fig. 8.2. The organic material from nano-Se^b showed fluorescence in the area zones associated with the proteins - tyrosine and tryptophan (I or II), and SMPs (IV) (Fig. 8.2a). In addition, peaks corresponding to fulvic-like acids (III) and humic-like (V) substances were evident. However, the relative fluorescence intensity of fulvic and humic-like substances (III and V) was low as compared to the fluorescence intensity of aromatic protein-like (areas I and II) and SMPs (IV). The 3D fluorescence spectra of organic material extracted from BSA stabilized nano-Se^c (Fig. 8.2b) exhibited a distinct peak (λ_{ex}~ 280 nm, λ_{em}~ 320 - 350 nm), which can be attributed to both the tryptophan and tyrosine residues of BSA (Ray et. al., 2012).

8.3.4.   Survival and hatching rate over time

The time courses of hatching of zebrafish embryos exposed to selenite and nano-Se are shown in Fig. 8.3. In the control, 93.75% of the embryos were hatched at 72 hpf. Compared with the controls, the hatching rates associated with selenite and nano-Se^b did not show any significant difference at nominal concentrations of < 1 mg. L^{-1} Se and all the embryos hatched. But, at concentrations above 2 mg. L^{-1} Se, hatching was delayed as compared to controls and 75% and 81% embryos were hatched in the presence of selenite and nano-Se^b, respectively. In contrast, the hatching rates of embryos exposed to nano-Se^{cs} were delayed compared to selenite and nano-Se^b even at the lowest concentration (0.2 mg. L^{-1}) and only 75% were hatched at 72 hpf. With an increase in nano-Se^{cs} concentration, hatching rate decreased further and only 25% embryos were hatched at 5 mg. L^{-1} at 72 hpf. The hatching rate in the presence of nano-Se^{cl} was not significantly affected as it did not show any substantial difference with the control at concentrations < 2 mg. L^{-1} Se. However, at 5 mg. L^{-1} concentration of nano-Se^{cl}, only 60% of embryos were hatched at 72 hpf.

From Fig. 8.4 it was evident that at 48 hpf, neither selenite nor any of nano-Se showed mortality even at the highest concentration (5 mg. L^{-1}) tested, but, mortality was evident after that. Nano-Se^{cs} showed higher mortality at 72 hpf than both of nano-Se^b and nano-Se^{cl}. Even at 1 mg. L^{-1} Se, nano-Se^{cs} caused 56.25% mortality at 72hpf, while neither selenite nor nano-Se^b showed mortality at this time point (Fig. 8.4). The % mortality of zebrafish embryos following exposure to selenite and different nano-Se at 120 hpf are given in Fig. 8.5. Exposure of embryos to different concentrations of selenite and different nano-Se suspensions caused a range of mortalities at 120 hpf. The LC50 values and 95% confidence interval (CI) of selenite and nano-Se^b were 0.55 (0.51 - 0.59) mg. L^{-1}, and 1.77 (1.59 - 1.98) mg. L^{-1}, respectively. It is evident that particle size did not have a significant effect on nano-Se^c toxicity as the LC50 values and 95% CI of nano-Se^{cs} and nano-Se^{cl} were comparable at 0.16 (0.15 - 0.17) mg. L^{-1} and 0.17 (0.15 - 0.19) mg. L^{-1} (Fig. 8.5). It also implies that nano-Se^b was 10 times less toxic than nano-Se^c independent of size. So, toxicity of selenium on zebrafish embryos was observed in the following order: nano-Se^b < selenite < nano-Se^{cs} ~ nano-Se^{cl}.

Table 8.1 Particle size distribution, zeta-potential of nano-Se^b, nano-Se^{cs} and nano-Se^{cl} after 1 h and 24 h of incubation in egg water

Nano-Se	Particles Size distribution (nm)		Zeta-potential (mV)	
	1 h	24 h	1 h	24 h
Nano-Se^b	100- 400	100-400	-28.5 ±5.19	-28.1 ± 4.86
Nano-Se^{cs}	25-90	25-100	-16.2 ± 3.16	-15.9 ± 3.97
Nano-Se^{cl}	50-250	50-250	-13.73 ± 3.57	-12.78 ± 4.11

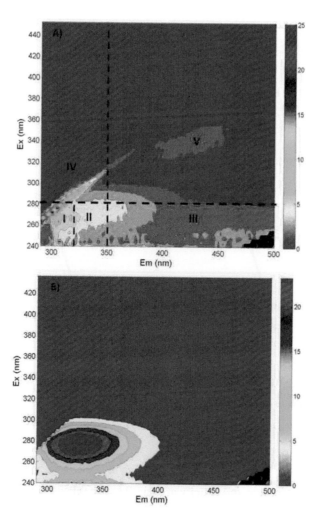

Fig. 8.2 Three dimensional EEM (Excitation Emission Matrix) fluorescence spectra of extracted organic material. (A) Natural organic material from nano-Se^b (B) BSA from nano-Se^{cs}. Region I and II - aromatic proteins; IV - soluble microbial by-products-like; III and V – Fulvic acid and humic-like substances

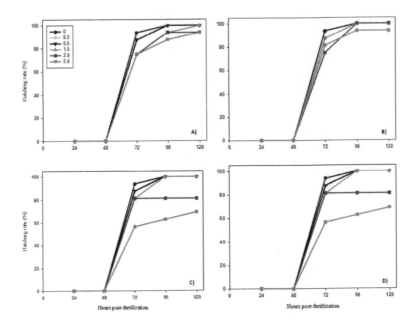

Fig. 8.3 Time course of hatching rate of surviving zebrafish embryos exposed to A) selenite, B) nano-Se^b, C) nano-Se^cl and D) nano-Se^cs suspensions from 24 h to 120 h post-fertilization (hpf). Legend symbols are inside the panel (A)

8.4. Discussion

8.4.1. Comparison of fate and toxicity of nano-Se^b with selenite and nano-Se^c

One major drawback of microbial bioremediation of selenium oxyanions is that a fraction of nano-Se^b formed in bioreactors will remain suspended in the treated effluents leaving the bioreactors, thus nano-Se^b entering the aquatic environments (Jain et. al., 2016; Lenz et. al., 2008; Soda et. al., 2011). Moreover, the selenium discharge criteria for aquatic life and the proposed toxicity thresholds are highly debated in recent times (Chapman, 1999; DeForest et. al., 1999). In spite of concerns of Se toxicity to aquatic ecosystems, there are no studies on the toxic effect of biogenic nano-Se on aquatic organisms. Hence, the fate and toxicity of nano-Se^b formed by anaerobic granular sludge biofilm were investigated on zebrafish embryos and compared with the fate and toxicity of selenite and chemogenic nano-Se.

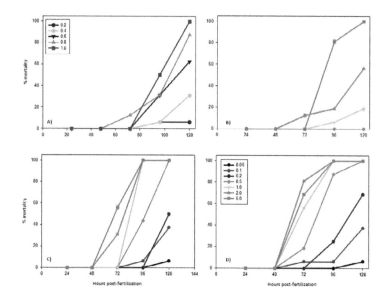

Fig. 8.4 Time course of mortality of zebrafish embryos exposed to A) selenite (legend symbols are inside), B) nano-Se[b], C) nano-Se[cl] and D) nano-Se[cs] suspensions from 24 h to 120 h post-fertilization (hpf). (B-D) - Legend symbols are inside the panel (D)

Under normal assay conditions, zebrafish embryos hatch at approximately 72 hpf (Hua et. al., 2014). Compared with the controls, the hatching rates associated with selenite and nano-Se[b] exposure were not significantly delayed during the 96 h exposure time (Fig. 8.3). However, increasing nano-Se[cs] and nano-Se[cl] concentrations had an inhibitory effect on the hatching rate and success (Fig. 8.3). It is possible that the nano-Se[c] suspension inhibited the secretion and activity of the hatching enzyme chorionase of zebrafish, as reported previously for other NPs like Cu and Zn NPs (Nel et. al., 2006). Surprisingly, nano-Se[cs] and nano-Se[cl] had a different effect on hatching, but they induced comparable mortality on zebrafish embryos. Although the exact mechanisms of hatching inhibition are not known, it is possible that the mechanisms of action of lethality and of hatching inhibition by NPs on zebrafish embryos are different (Hua et. al., 2014).

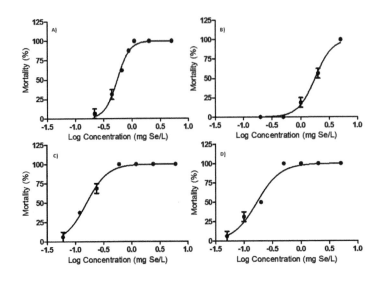

Fig. 8.5 Dose-response curves of mortality of zebrafish embryos at 120 h postfertilization following exposure to A) selenite, B) nano-Se[b], C) nano-Se[cl] and D) nano-Se[cs]

Based on the LC50 values (Fig. 8.5), nano-Se[b] was 3.2-fold less toxic than selenite and 10-fold less toxic than nano-Se[c]. The result was in contrast to Zhang et al. (2001; 2005) and Wang et al. (2007), where it was reported that chemogenic nano-Se had 7-fold and 3.5-fold lower acute toxicity than sodium selenite and L-selenomethionine, respectively, in Kunming mice. However, Li et al. (2008) reported that chemogenic nano-Se particles were more toxic than selenite in medaka fish. Recently, Shakibaie et al. (2013) also reported that biogenic nano-Se synthesized by *Bacillus* sp. MSh-1 was less toxic than chemogenic nano-Se and SeO2 in mice. The discrepancy in the toxicities of nano-Se can result from the differences in nano-Se particle size, surface stabilizing agents, medium composition, and test organisms. The relative contribution to toxicity of nano-Se$_{particle}$ and nano-Se$_{ion}$ of all three nano-Se to zebrafish embryos using response addition and concentration addition was calculated (Equation 2). As the dissolution was marginal (Fig. S8.3, Appendix 1), calculations showed that toxicity caused by the nano-Se$_{ion}$ was negligible. The nano-Se$_{particle}$ of all three nano-Se were thus found to be the main factor causing toxicity to zebrafish embryos.

Interestingly, nano-Se[c] with two different sizes such as nano-Se[cs] (25-90 nm) and nano-Se[cl] (50-250 nm) showed no significant (P>0.05) difference in LC50 values (Fig. 8.5) and dissolution kinetics (Fig. S8.3, Appendix 1), suggesting that particle size has no effect on nano-Se fate and toxicity. It was reported that nano-Se has no size effect in the induction of Se-containing enzymes like glutathione peroxidase in mice suggesting that bioavailability (uptake) plays a critical role in determining the toxicity of nano-Se (Zhang et. al., 2004). But, higher zeta-potential values (Table 8.1) and lower dissolution kinetics of nano-Se[b] (Fig. S8.3, Appendix 1) compared to nano-Se[c] support the contention that they both behave differently in terms of physicochemical stability and toxicity and the surface modification (capping agent) probably plays an important role in determining the fate of nano-Se (Misra et. al., 2012; Gunsolus et. al., 2015).

Recently, Fang et al. (2015) also reported that aggregation and hydrodynamic diameter of TiO$_2$ nanoparticles in suspensions are not the deciding factors for the toxicity of TiO$_2$ nanoparticles on zebrafish (*Danio rerio*), rather it was the humic acid which decreases the bioavailability along with suppression of oxidative stress. Liu and Hurt (2010) also showed that adsorption of NOM inhibited the ionic silver release from nano-Ag, while Moreau et al. (2007) showed that microbially-derived extracellular proteins changes the transport properties of biogenic ZnS nanoparticles by retarding their dispersal. However, studies on the effect of organic matter or EPS as surface coating on nano-Se is limited and more studies are needed to link the surface coating of nano-Se with their properties including their stability and dissolution in order to fully understand their environmental transformation, transport and fate.

8.4.2.   Effect of EPS on nano-Se toxicity

Nano-Se[b] and nano-Se[c] showed very contrasting toxic effects on zebrafish embryos development. The differences in toxicity can be attributed to the difference in surface stabilizing agents of nano-Se[b] and nano-Se[c]. Nano-Se[c] is stabilized by a single protein, BSA, while nano-Se[b] is stabilized by a complex mixture of biomolecules present on the surface of nano-Se[b] possibly originating from the microorganisms present in the anaerobic granular sludge biofilms (Jain et al., 2015). The presence of characteristic peaks for proteins, SMPs and fulvic and humic-like acids in nano-Se[b] provided evidence of the complex nature of the EPS in stabilizing nano-Se[b] (Fig. 8.2). A recent study showed that EPS extracted from anaerobic granular sludge directs the surface characteristics of chemogenic nano-Se similar to the biogenic nano-Se formed by anaerobic granular sludge (Jain et al., 2015).

Recently, Bondarenko et al. (2016) showed that the addition of a surface coating of levan, a fructose-composed biopolymer of bacterial origin, significantly reduced the toxic effects of Se-NPs in an in vitro assay on the human cell line Caco-2 cell line (colorectal adenocarcinomatous tissue of the human colon). The physiochemical stability of nano-Se[b] formed by biological route are higher than that of nano-Se[c] obtained by chemical synthetic methods (Oremland et al., 2004) and the presence of proteins in EPS possibly help in stabilizing nano-Se[b] (Sharma et al., 2014). Interestingly, the presence of recalcitrant components of EPS such as humic-like acids can reduce the bioavailability of contaminants to organisms due to the increased electrostatic repulsion (Lundqvist et al., 2010; Tang et al., 2015). A decrease in the selenium uptake by plant roots due to interactions between selenite and fulvic acid was reported (Wang et al., 1996) and it was suggested that association of selenium with organic matter in aquatic environments may play an important role in the mobility and bioavailability of selenium.

There are clear indications that the presence on humic acid like components in surface coating by EPS limits the bioavailability of nanoparticles and attenuates their toxicity. Gao et al. (2012) showed that by increasing humic acids concentration in surface coating, the toxicity of silver nanoparticles to *Daphina magna* was decreased. Moreover, presence of humic acids inhibit the generation of intracellular reactive oxygen species (ROS) and cellular lipid peroxidation which could be responsible for the lower toxicity of nanoparticles coated with natural organic matter (Lin et al., 2012; Yin et al., 2010). In contrast, BSA is commonly used to increase the bioavailability of NPs in *in-vivo* or *in-vitro* drug delivery (Kim et al., 2009). While BSA increases the colloidal stability of nanoparticles, it also increases the bioavailability of nano-Se[c], thereby making the particles to be more toxic.

Furthermore, from Table 8.1, it is clear that the ζ-potential is much higher in nano-Se[b] (-28.5 ± 5 mV) than nano-Se[cs] (-16 ± 3.1 mV) and nano-Se[cl] (-13.7 ± 3.6 mV). Thus, there is a higher degree of repulsion between the nano-Se[b] and zebrafish embryo due to electrostatic repulsion. But due to the lower negative potential in nano-Se[cs] and nano-Se[cl], the electrostatic repulsion is reduced which possibly increases the chances of the interaction of nano-Se[c] with embryos and results in higher toxicity (El Badawy et al., 2011). Based on the data, it is contended that the presence of EPS increases the colloidal stability of nano-Se and prevents interaction of biogenic Se particles and embryos due to electrostatic repulsion (Buchs et al., 2013; Lin et al.,

2012). This leads to lower bioavailability (uptake) and lower toxicity of nano-Se[b] suggesting that EPS play a critical role in determining the fate and toxicity of biogenic Se nanoparticles.

8.4.3.   Environmental implications

Although full scale biological systems are available for treating selenium wastewaters, due to colloidal properties a fraction of biogenic nano-Se will remain in the treated waters, thus leaving the bioreactors with the effluent (Lenz et al., 2008; Soda et al., 2011). The present study provided experimental evidence for the first time that nano-Se[b] formed by the biofilms in the bioreactors is comparatively much less toxic to aquatic organisms than selenite and nano-Se[c]. This has implications in alleviating the concerns of nano-Se[b] leaving the bioreactors and possibly adoption of new procedures and guidelines on selenium toxicity evaluation.

Moreover, Se is a semiconductor and scarce element (Nancharaiah et al., 2016) used in many applications like production of photovoltaic cells, rectifiers and Se coated cylindrical drums for xerography (Poborchii et al., 1998) and there has been an increased interest in synthesizing selenium nanoparticles using microorganisms as inexpensive "green" catalysts (Nancharaiah & Lens, 2015b). While the rapid development in the synthesis and commercialization of nano-Se along with other nanomaterials is imminent, these trends may pose hazards to ecosystem well-being and human health. The present study suggests that more emphasis should be given to the use of biological synthesis for commercial production of nano-Se. The toxicity profiles of nano-Se[b] and nano-Se[c] also demonstrate the importance of selecting the surface coating materials while considering safer nano-Se for environmental purposes, or antimicrobial nano-Se for biomedical applications. However, additional studies are warranted to investigate the bioavailability, bioaccumulation, stability and toxicity of accumulated nano-Se[b] or surface modified nano-Se[c].

8.5. Conclusions

This study provided evidence that biogenic nano-Se are 3.2-fold less toxic than selenite and 10-fold less toxic than chemogenic nano-Se to zebrafish. In addition, smaller sized and larger sized BSA capped chemogenic nano-Se showed comparable LC50 values on zebra fish embryos. LC50 values with 95% CI for selenite, nano-Se[b], nano-Se[cs] and nano-Se[cl] were 0.55 (0.51-0.59) mg. L[-1], 1.77 (1.593-1.983) mg. L[-1], 0.1593 (0.1444-0.1756) mg. L[-1] and 0.1703

(0.1457-0.1991) mg. L^{-1}, respectively. Biogenic nano-Se was found to be more stable in terms of dissolution kinetics compared to chemogenic nano-Se. 3D-EEM analysis of organic matter associated with biogenic nano-Se showed distinct peaks for proteins, soluble microbial products and humic acids typical for EPS matrix of biofilms.

Acknowledgements

The authors declare that there are no conflicts of interest. This research was supported through the Erasmus Mundus Joint Doctorate Environmental Technologies for Contaminated Solids, Soils, and Sediments (ETeCoS^3) (FPA n^0 2010-0009) and BioMatch (project No. 103922), funded by the European Commission Marie Curie International Incoming Fellowship (MC-IIF). The authors would like to thank Dr. Wouter Veneman, Dr. Willie Peijnenburg and Dr. Martina Vijver (Leiden University) for the collaboration. The authors would like to thank Gerda Lamers (Leiden University) for the technical help of transmission electron microscopy.

References

Backhaus, T., Scholze, M, Grimme, L.H. 2000. The single substance and mixture toxicity of quinolones to the bioluminescent bacterium *Vibrio fischeri*. *Aquat Toxicol.*, **49**, 49-61.

Baker, A., Lamont-Black, J. 2001. Fluorescence of dissolved organic matter as natural tracer of ground water. *Ground Water.*, **39**, 745-750.

Bhatia, D., Bourven, I., Simon, S., Bordasa, F., van Hullebusch, E.D., Rossano, S., Lens, P.N.L., Guibaud, G. 2013. Fluorescence detection to determine proteins and humic-like substances fingerprints of exopolymeric substances (EPS) from biological sludges performed by size exclusion chromatography (SEC). *Bioresour Technol.*, **131**, 159-165.

Bondarenko, O.M., Ivask, A., Kahru, A., Vija, H., Titma, T., Visnapuu, M., Joost, U., Pudova, K., Visnapuu, T., Alamäe, T. 2016. Bacterial polysaccharide levan as stabilizing, non-toxic and functional coating material for microelement-nanoparticles. *Carbohydr Polym.*, **20**, 710-720.

Buchs, B., Evangelou, M.W., Winkel, L.H., Lenz, M. 2013. Colloidal properties of nanoparticular biogenic selenium govern environmental fate and bioremediation effectiveness. *Environ Sci Technol.*, **47**, 2401-2407.

Chapman, P.M. 1999. Selenium-a potential time bomb or just another contaminant? *Hum Ecol Risk Assess.*, **5**, 1123-1138.

deForest, D.K., Brix, K.V., Adams, W.J. 1999. Critical review of proposed residuebased selenium toxicity thresholds for freshwater fish. *Hum Ecol Risk Assess.*, **5**, 1187-1228.

Dennis, L.A. 2004. Aquatic selenium pollution is a global environmental safety issue. *Ecotoxicol Environ Saf.*, **59**, 44-56.

El Badawy, A.M., Silva, R.G., Morris, B., Scheckel, K.G., Suidan, M.T., Tolaymat, T.M. 2011. Surface Charge-Dependent Toxicity of Silver Nanoparticles. *Environ Sci Technol.*, **45**, 283-287.

Fang, T., Yu, L.P., Zhang, W.C., Bao, S.P. 2015. Effects of humic acid and ionic strength on TiO$_2$ nanoparticles sublethal toxicity to zebrafish. *Ecotoxicology*, **24**, 2054-2066

Gao, J., Powers, K., Wang, Y., Zhou, H., Roberts, S.M., Moudgil, B.M., Koopman, B., Barber, D.S. 2012. Influence of Suwannee River humic acid on particle properties and toxicity of silver nanoparticles. *Chemosphere*, **89**, 96-101.

Gunsolus, I.L., Mousavi, M.P., Hussein, K., Bühlmann, P., Haynes, C.L. 2015. Effects of Humic and Fulvic Acids on Silver Nanoparticle Stability, Dissolution, and Toxicity. *Environ Sci Technol.*, **49**, 8078-8086.

Hua, J., Vijver, M.G., Ahmad, F., Richardson, M.K., Peijnenburg, W.J. 2014. Toxicity of different-sized copper nano- and submicron particles and their shed copper ions to zebrafish embryos. *Environ Toxicol Chem.*, **33**, 1774-1782

Jain, R., Jordan, N., Schild, D., van Hullebusch, E.D., Weiss, S., Franzen, C., Farges, F., Hübner, R., Lens, P.N.L. 2015. Extracellular polymeric substances govern the surface charge of biogenic elemental selenium nanoparticles. *Environ Sci Technol.*, **49**, 1713-1720.

Dessi, P., Jain, R., Singh, S., Seder-Colomina, M., van Hullebusch, E.D., Ahammad, S.Z., Rene, E.R., Carucci, A., Lens, P.N.L. 2016. Effect of temperature on selenium removal from wastewater by UASB. *Water Res.*, **94**, 146-154.

Kim, B.S., Oh, J.M., Kim, K.S., Seo, K.S., Cho, J.S., Khang, G., Lee, H.B., Park, K., Kim, MS. 2009. BSA-FITC-loaded microcapsules for in vivo delivery. *Biomaterials*, **30**, 902-909.

Leenheer, J.A., Croué, J.P. 2003. Characterizing aquatic dissolved organic matter. *Environ Sci Technol.*, **37**, 18-26.

Lenz, M., Lens, P.N.L. 2009. The essential toxin, The changing perception of selenium in environmental sciences. *Sci Total Environ.*, **407**, 3620-3633.

Lenz, M., van Hullebusch, E.D., Hommes, G., Corvini, P.F.X., Lens, P.N.L. 2008. Selenate removal in methanogenic and sulfate-reducing upflow anaerobic sludge bed reactors. *Water Res.,* **42**, 2184-2192.

Li, H.C., Zhang, J.S., Wang, T., Luo, W.R., Zhou, Q.F, Jiang, G.B. 2008. Elemental selenium particles at nanosize (Nano-Se) are more toxic to medaka (Oryzias latipes) as a consequence of hyperaccumulation of selenium, A comparison with sodium selenite. *Aquat Toxicol.,* **89**, 251-256.

Lin, D., Ji, J., Long, Z., Yang, K., Wu, F. 2012. The influence of dissolved and surface-bound humic acid on the toxicity of TiO_2 nanoparticles to *Chlorella* sp. *Water Res.,* **46**, 4477-4487.

Liu, H., Fang, H.H. 2002. Extraction of extracellular polymeric substances (EPS) of sludges. *J Biotechnol.,* **95**, 249-256.

Liu, J.Y., Hurt, R.H. 2010. Ion release kinetics and particle persistence in aqueous nano-silver colloids. *Environ Sci Technol.,* **44**, 2169–2175.

Lohr, J.L., Yost, H.J. 2000. Vertebrate model system in the study of early heart development, Xenopus and zebrafish. *Am J Med Genet.,* **97**.

Lundqvist, A., Bertilsson, S., Goedkoop, W. 2010. Effects of extracellular polymeric and humic substances on chlorpyrifos bioavailability to *Chironomus riparius*. *Ecotoxicology*, **19**, 614-622.

Luoma, S.N., Johns, C., Fisher, N.S., Steinberg, N.A., Oremland, R.S., Reinfelder, J.R. 2000. Determination of selenium bioavailability to a benthic bivalve from particulate and solute pathways. *Environ Sci Technol.,* **26**, 485-491.

Maeng, S.K., Sharma, S.K., Abel, C.D., Magic-Knezev, A., Song, K.G., Amy, G.L. 2012. Effects of effluent organic matter characteristics on the removal of bulk organic matter and selected pharmaceutically active compounds during managed aquifer recharge, Column study. *J Contam Hydrol.,* **140-141**, 139-149.

Mal, J., Nancharaiah, Y.V., van Hullebusch, E.D., Lens, P.N.L. 2016. Effect of heavy metal co-contaminants on selenite bioreduction by anaerobic granular sludge. *Bioresour Technol.,* **206**, 1-8.

Misra, S.K., Dybowska, A., Berhanu, D., Luoma, S.N., Valsami-Jones, E. 2012. The complexity of nanoparticle dissolution and its importance in nanotoxicological studies. *Sci Total Environ.,* **438**, 225-232.

Moreau, J.W., Weber, P.K., Martin, M.C., Gilbert, B., Hutcheon, I.D., Banfield, J.F. 2007. Extracellular proteins limit the dispersal of biogenic nanoparticles. *Science,* **316,** 1600-1603.

Nancharaiah, Y.V., Lens, P.N.L. 2015a. Ecology and biotechnology of selenium-respiring bacteria. *Microbiol Mol Biol Rev.,* **79,** 61-80.

Nancharaiah, Y.V., Lens, P.N.L. 2015b. Selenium biomineralization for biotechnological applications. *Trends Biotechnol.,* **33,** 323-330.

Nancharaiah, Y.V., Venkata S.M., Lens, P.N.L. 2016. Biological and bioelectrochemical recovery of critical and scarce metals. *Trends Biotechnol.,* **34,** 137-155.

Nel, A., Xia, T., Madler, L., Li, N. 2006. Toxic potential of materials at the nanolevel. *Science,* **311,** 622-627.

Oremland, R.S., Herbel, M.J., Blum, J.S., Langley, S., Beveridge, T.J., Ajayan, P.M., Sutto, T. 2004. Structuraland spectral features of selenium nanospheres produced by Se-respiring bacteria. *Appl Environ Microbiol.,* **70,** 52-60.

Organisation for Economic Co-operation and Development. 2011. Validation report (phase 1) for the zebrafish embryo toxicity test, part I. Series on Testing and Assessment No. 157. ENV/JM/MONO (2011) 37 Paris, France.

Pearce, C.I., Pattrick, R.A.D., Law, N., Charnock, J.M., Coker, V.S., Fellowes, W.J., Oremland, S.R., Lloyd, R.J. 2009. Investigating different mechanisms for biogenic selenite transformations, *Geobacter sulfurreducens, Shewanella oneidensis* and *Veillonella atypica. Environ Technol.,* **30,** 1313-1326.

Poborchii, V.V., Kolobov, A.V., Tanaka, K. 1998. An *in-situ* Raman study of polarization-dependent photocrystallization in amorphous selenium films. *Appl Phys Lett.,* **72,** 1167-169.

Ray, D., Paul, B.K., Guchhait, N. 2012. Effect of Biological Confinement on the Photophysics and Dynamics of a ProtonTransfer Phototautomer, An Exploration of Excitation and Emission WavelengthDependent Photophysics of the Protein-Bound Drug. *Phys Chem Chem Phys.,* **14,** 12182-12192.

Rayman, M.P. 2006. The importance of selenium to human health. *Lancet,* **356,** 23-241.

Scheinost, A.C., Kirsch, R., Banerjee, D., Fernandez-Martinez, A., Zaenker, H., Funke, H., Charlet, L. 2008. X-ray absorption spectroscopy investigation of selenite reduction by Fe(II)-bearing minerals. *J Contam Hydrol.,* **102,** 228-245.

Schlekat, C.E., Dowdle, P.R., Lee, B.G., Luoma, S.N., Oremland, R.S. 2000. Bioavailability of particle-associated Se to the bivalvePotamocorbula amurensis. *Environ Sci Technol.*, **30**, 4504-4510.

Shakibaie, M., Shahverdi, A.R., Faramarzi, M.A., Hassanzadeh, G.R., Rahimi, H.R., Sabzevari, O. 2013. Acute and subacute toxicity of novel biogenic selenium nanoparticles in mice. *Pharm Biol.*, **51**, 58-63.

Sharma, V.K., Siskova, K.M., Zboril, R., Gardea-Torresdey, J.L. 2014. Organic-coated silver nanoparticles in biological and environmental conditions, Fate, stability and toxicity. *Adv Colloid Interface Sci.*, **204**, 15-34.

Soda, S., Kashiwa, M., Kagami, T., Kuroda, M., Yamashita, M., Ike, M. 2011. Laboratory-scale bioreactors for soluble selenium removal from selenium refinery wastewater using anaerobic sludge. *Desalination,* **279**, 433-438.

Srivastava, N., Mukhopadhyay, M. 2013. Biosynthesis and structural characterization of selenium nanoparticles mediated by *Zooglea ramigera*. *Powder Technol.*, **244**, 26-29.

Tang, Y., Li, S., Lu, Y., Li, Q., Yu, S. 2015. The influence of humic acid on the toxicity of nano-ZnO and Zn^{2+} to the *Anabaena* sp. *Environ Toxicol.*, **30**, 895-903.

Wang, H., Zhang, J., Yu, H. 2007. Elemental selenium at nano size possesses lower toxicity without compromising the fundamental effect on selenoenzymes, Comparison with selenomethionine in mice. *Free Radic Biol Med.*, **42**, 1524-1533.

Wang, Z.J.; Xu, Y., Peng, A. 1996. Influences of fulvic acid on bioavailability and toxicity. *Biol Trace Elem Res.,* **55**, 147-162.

Winkel, L.H., Johnson, C.A., Lenz, M., Grundl, T., Leupin, O.X., Amini, M., Charlet, L. 2012. Environmental selenium research, from microscopic processes to global understanding. *Environ Sci Technol.*, **46**, 571-579.

Yin, H., Casey, P.S., McCall, M.J., Fenech, M. 2010. Effects of surface chemistry on cytotoxicity, genotoxicity, and the generation of reactive oxygen species induced by ZnO nanoparticles. *Langmuir,* **26**, 15399-15408.

Zhang, J., Wang, H., Bao, Y., Zhang, L. 2004. Nano red elemental selenium has no size effect in the induction of seleno-enzymes in both cultured cells and mice. *Life Sci.* 75, 237-244.

Zhang, J.S., Gao, X.Y., Zhang, L.D., Bao, Y.P. 2001. Biological effects of a nano red elemental selenium. *Biofactors.* **15**, 27-38.

Zhang, J.S., Wang, H.L., Yan, X.X., Zhang, L.D. 2005. Comparison of short-term toxicity between Nano-Se and selenite in mice. *Life Sci.,* **76**, 1099–1109.

Zhang. W., Chen, Z., Liu, H., Zhang, L., Gao, P., Li, D. 2011. Biosynthesis and structural characteristics of selenium nanoparticles by *Pseudomonas alcaliphila*. *Colloids Surf B Biointerfaces.*, **88**, 196-201.

Zhang, Y.Q., Moore, J.N. 1996. Selenium fractionation and speciation in a wetland system. *Environ Sci Technol.*, **30**, 2613-2619.

Zhou, M., Liu, B., Lv, C., Chen, Z., Shen, J. 2014. Rapid synthesis of NADPH responsive CdSe quantum dots from selenium nanoparticles. *RSC Adv.*, **4**, 61133-61136.

Chapter 8

CHAPTER 9

Discussion, conclusions and perspectives

9.1. General discussion

Recent years have seen a growing interest in the application of chalcogenide (e.g. Se, Te and CdSe) nanoparticles (NPs) in various industrial sectors including energy, food, cosmetic, steel and glass (Mal et al., 2016d). Microbial reduction of chalcogen oxyanions is a promising technology for bioremediation and treatment of Se/Te wastewaters and remains a matter of interest and discussion. This thesis aimed to develop a microbial synthesis process to produce chalcogenide NPs from synthetic wastewater by combining bioremediation and biorecovery of Se/Te in the form of Se/Te chalcogenide nanoparticles. The sub-objectives were: biogenic Se and Te NPs production (**Chapters 3, 4, 5, and 7**), biosynthesis of CdSe NPs (**Chapters 5, 6**), studying the fate and toxicity of biogenic Se NPs (**Chapter 8**). This **Chapter 9** provides a valuable insight into the current state of the microbial synthesis of chalcogenide NPs, indicating the current challenges and future perspectives of these technologies based on the findings of the present thesis (Fig. 9.1). All these results can certainly be exploited to develop microbial synthesis of NPs of other metal (e.g. Se-Te, CuSe, and CuTe) in the near future.

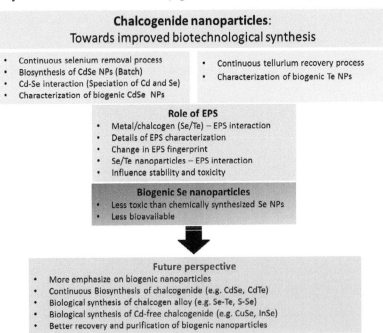

Fig. 9.1 Biogenic chalcogenide nanoparticles - overview of current finding and future perspectives

9.1.1.   Microbial synthesis of selenium-based chalcogenide nanoparticles

Biological treatment of Se-contaminated wastewaters for practical applications may have important limitations because of the presence of NH_4^+/NO_3^- or SO_4^{2-} or by the presence of heavy metals (e.g. Cd, Zn, and Pb) co-contaminants. **Chapter 3** demonstrated that both selenate and ammonium-nitrogen could be removed via bioreduction and biological nitrification-denitrification, respectively, by using activated sludge in a SBR operation. Using the SBR configuration with alternating anaerobic and aerobic phases in the cycle period, high selenium removal efficiencies were achieved. The highest ammonium-nitrogen removal efficiency achieved was 98 % while total nitrogen (TN) removal was 75 %. Interestingly, with the increase in the initial ammonium concentration, both the Se and TN removal efficiency increased during batch experiments, suggesting that further studies are required on the ammonium/selenate ratio to increase the treatment efficiencies in the SBR. The study also suggested the importance of the time duration of each of the anoxic or aerobic phases which need to be further optimized to achieve optimum reactor performance.

The effect of heavy metal co-contaminants on selenite bioreduction by anaerobic granular sludge is described in **Chapter 4**. Among Zn, Pd and Cd, only Cd shows inhibition on selenite bioreduction at concentrations >150 mg. L^{-1}. Based on the results, it is clear that the removal of heavy metals is mainly driven by biosorption onto the granular sludge (Yuan et al., 2009) and adsorption of heavy metals onto Se(0) nanoparticles. Subsequent reduction of Se(0) forms HSe^- which precipitates with heavy metals to form metal selenides. The formation of Se(0) or HSe^- was influenced by the type and concentration of heavy metals (Mal et al., 2016a). Hence, it is suggested that selection of heavy metals and optimization of its initial concentrations are highly important for microbial metal selenide synthesis. It was evident that the concentration of Se (0) and HSe^- in the liquid phase was relatively higher in the presence of Cd as compared to Pb and Zn.

Based on the results of **Chapter 4**, microbial synthesis of CdSe by utilizing a selenium reducing microbial community was explored in **Chapter 5.** The microbial community in anaerobic granular sludge was enriched for 300 days in the presence Cd(II) and Se(IV). The microbial community was successfully enriched to reduce selenite up to selenide in the presence of Cd and depending on the size and surface properties, the formation of CdSe was observed both in the granular sludge and in the aqueous phase. Further speciation studies using X-ray absorption spectroscopic techniques e.g. XANES and EXAFS to evaluate the details of the chemical environment of Cd and Se are required (Mal et al., 2016a). In **Chapter 5**, it was revealed that also Raman spectroscopy and XPS can be used successfully for studying the speciation of cadmium and selenium.

The XPS and Raman spectroscopy indicates that CdSe is present in the granular sludge as well as in the aqueous phase. Interestingly, a CdS_xSe_{1-x} ternary structure might have formed around the CdSe core on the surface of the granular sludge (Dzhagan et al., 2013). It is possible that Cd ions present on the surface of the CdSe core are bonded with the S ions (sulfhydryl group; -SH group), mainly present in the EPS originating from an anaerobic granular sludge biofilm (Raj et al., 2016). In the aqueous phase, the signal from CdSe was stronger without any impurities like CdS formation, which was highly significant as the production of CdSe in the aqueous phase is more desirable due to easier separation and recovery. However, another limitation was the presence of trigonal Se along with CdSe resulting in poor crystallinity of CdSe, which suggests that CdSe needs to be separated and purified from Se nanoparticles.

9.1.2. Microbial synthesis of tellurium-based chalcogenide

Recently extensive research has been conducted in the development of new Te-based materials such as fluorescent CdTe quantum dots due to its excellent thermal, optical and electrical properties (Mal et al., 2016d). But being one of the least abundant elements in the lithosphere, development of new technologies are absolutely essential for the recovery of Te from mining waste streams and from its end-use applications to ensure its availability (Ramos-Ruiz et al., 2016). Therefore, the feasibility of laboratory-scale upflow anaerobic granular sludge bed (UASB) reactors inoculated with anaerobic granular sludge for continuous bioreduction of Te(IV) and recovery of biogenic Te(0) was explored in **Chapter 7**. Bioreduction of Te(VI) and deposition of biogenic Te(0) on anaerobic granular sludge was the main mechanism of Te(IV) removal. The result shows that the microorganisms present in the anaerobic granular sludge readily reduced Te(IV) and did not require any special treatment like enrichment or acclimatization for Te(IV) reduction.

Wash-out of Te(0) particles in the reactor effluent could be minimized by using granular sludge and it was evident that the majority of the Te was entrapped within the granular sludge.

Moreover, Te(0) could be easily recovered by the centrifugation method along with EPS which makes the process more attractive and simpler. 74 – 78% of the biomass associated Te(0) was recovered by centrifugation (10000 × g for 20 mins), indicating that the biogenic Te(0) was associated predominantly with loosely bound extracellular polymeric substances (EPS).

Low-magnification scanning electron microscopy (SEM) coupled with energy dispersive X-ray (EDX) imaging confirms the presence of Te(0) nanocrystals on the surface of the granular sludge. Characterization of the granular sludge by Raman spectroscopy further confirms the deposition of Te(0) onto the granular sludge. Two characteristic vibration peaks at 121.9 and 140.5 cm^{-1} of Te confirm the presence of metallic Te(0) (Bonificio & Clarke, 2014; Li et al., 2014). The presence of the characteristic diffraction peaks of crystalline Te(0) in XRD validates the evidence of biosynthesis of Te(0) nanocrystals. All peaks in this pattern could be indexed to hexagonal tellurium (Li et al., 2014; Xue et al., 2012). Adding to that, no other characteristic peaks of impurities such as TeO_2 and K_2TeO_3 are detected either in Raman spectroscopy or XRD analysis. The present study (**Chapter 7**) shows that Raman spectroscopy and XRD can be used successfully for characterizing biogenic Te(0) as it is very easy to unambiguously distinguish the presence of metallic tellurium.

9.1.3. EPS characterization – understanding the change in composition and fingerprints of EPS during Se/Te-containing wastewater treatment

EPS are a complex mixture of high molecular weight macromolecules comprising mainly of proteins, carbohydrates, and humic-like substances (Bhatia et al., 2013). EPS are usually the first barrier of microbial cells and play an important role in biosorption and redox cycling of metals in the environment (Li & Yu, 2014; Tourney & Ngwenya, 2014). Several reports on the significant contribution of EPS on biologically-induced mineralization of calcium carbonate (Wei et al., 2015) or on both adsorption and reduction of metal(loid) like silver, cadmium sulfide and uranium are well documented (Li et al., 2016; Raj et al., 2016; Stylo et al., 2013). In **Chapters 6 and 7,** a detailed characterization of EPS is given to understand the change in EPS composition and/or fingerprint due to treatment of Se/Te-containing wastewater.

The increase in protein/polysaccharide (PN/PS) ratio in the EPS samples after the enrichment of granular sludge in the presence of Cd and Se, while the amount of humic (HS)-like substances reduced indicate that PN play a more important role in CdSe synthesis as well as protecting the granules from Cd and Se as well as the CdSe NPs (**Chapter 6**). All three components, i.e. PN, PS and HS-like substances, increased after Te-containing wastewater treatment for 70 days (**Chapter 7**). The increase in PN (40%) content in the EPS after the treatment of Se or Te containing wastewater was more significant than the PS (11%) and HS (16%), suggesting that the protein fraction may play a more important role in Te-containing wastewater treatment and biosynthesis of Te(0) similar to CdSe. Both studies show that extracellular proteins might be more involved than polysaccharides in treatment of Se or Te-containing wastewater.

The present study (**Chapters 6 and 7**) shows that EEM can be a useful tool to elucidate the change in chemical composition of EPS samples as a spectral fingerprint technology. Particularly detailed analysis of peak location and fluorescence intensity could be used for quantitative analysis. A red shift is generally associated with the presence of some functional groups like carbonyl, hydroxyl, amino groups and carboxyl and a blue shift indicates the break-up of the large molecules into smaller fragments (Zhu et al., 2015). These observed shifts in fluorescence peaks could also be due to the formation of biogenic CdSe-EPS and/or nano-Se-EPS complexes in microbial CdSe synthesis or could be due to the formation of Te nanocrystals-EPS complexes in biogenic Te(0) recovery systems (Wang et al., 2016).

In **Chapter 7,** it is demonstrated that size exclusion chromatography (SEC) coupled to a fluorescence detector for protein-like (221/350 nm) and humic-like (345/443 nm) substances can give access to valuable information about the fingerprints and/or the distribution of apparent molecular weight of EPS present in a sludge or biofilm. All these results reveal that there was a distinct change in chemical composition and/or in the fingerprint of EPS extracted from anaerobic granular sludge after the microbial synthesis of CdSe or Te(0). Moreover, the presence of biogenic CdSe or Te(0) nanoparticles associated with EPS suggests that the role of EPS in microbial synthesis of biogenic chalcogenide nanoparticles should not be ignored. Establishing the potential role of EPS in reduction and immobilization of the chalcogens will certainly add an important missing element in our understanding of chalcogen cycling, particularly in biofilms.

9.2. Synthesis of biogenic chalcogen alloys (e.g. Se/Te)

Se/Te alloys have unique semiconducting and optical properties with potential uses in optoelectronics such as light emitting diode (LED) (Tripathi et al., 2009). Moreover, Se/Te alloys have enhanced properties compared to single Te and Se nanomaterials, including an unusual temperature dependence, electrical resistivity and magnetoresistance (Sadtler et al., 2013; Sridharan et al., 2013). Simultaneous existence of selenium and tellurium occurs in copper and sulfur bearing ores, as well as in metal refinery wastewater (Soda et al., 2011). Hence, the feasibility of a laboratory-scale upflow anaerobic granular sludge bed (UASB) reactor for combining simultaneous bioreduction of selenium and tellurium and recovery of Se/Te was evaluated, similar to the continuous Te(0) recovery as described in **Chapter 7**. Anaerobic granular sludge was used as inoculum and lactate as electron donor at an organic loading rate of 0.6 g COD L^{-1} D^{-1}. Selenite and tellurite were provided in the concentration of 0.05 mM each, i.e. 4 mg Se. L^{-1} and 6 mg Te. L^{-1} and removal efficiencies of >97% were achieved for both.

The majority of the Se(IV) and Te(VI) was reduced Se(0) and Te(0), respectively, and was associated with the biomass as confirmed by SEM-EDX (Fig. 9.2), and Raman spectroscopy (Fig. 9.3). Noticeable peaks for carbon, oxygen, nitrogen along with Se(0) and Te(0) can be attributed to the deposition of biogenic Se(0) and Te(0) on the surface granular sludge. However, it was not possible to say whether there was any formation of Se-Te alloy nanocrystals or not. Characterization of the granular sludge by Raman spectroscopy further confirms the deposition of Se(0) and Te(0) (Fig. 9.3). Two characteristic vibration peaks at 122 and 140 cm^{-1}, are attributed to metallic Te(0), while the peak at 237 cm^{-1} was assigned to trigonal-Se (t-Se) (Iovu et al., 2005).

However, broadening of the t-Se peak at 237 cm^{-1} and split into several narrow distinct peaks suggest the inclusion of Te in the Se ring structure. The peak at ~ 204 cm^{-1} which has been assigned to vibration of neighboring amorphous Se-Te bonds, while the crystalline Se-Te peak generally appears at ~ 170 cm^{-1} (Iovu et al., 2005; Tang et al., 2015). Raman spectroscopy clearly indicates the formation of new structural units consisting of both Te and Se atoms. This result shows that simultaneous bioreduction of Se and Te-oxyanion is a promising technology for combined removal of both metalloids from polluted effluents and microbial synthesis of Se/Te alloys. It is an ongoing study and more studies are required including TEM and XRD for detailed characterization of Se/Te alloys to know its other properties like size and crystallinity.

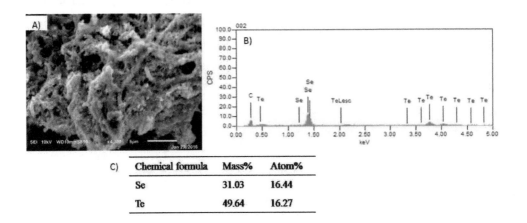

C)	Chemical formula	Mass%	Atom%
	Se	31.03	16.44
	Te	49.64	16.27

Fig. 9.2 A) SEM image of granular sludge taken from the UASB reactor after 48 days of Se/Te-containing wastewater treatment, B) EDX shows the presence of Se(0) and Te(0) corresponding to the samples shown in the SEM image. C) Chemical composition (%) of Se(0) and Te(0) elements detected on the surface of a granular sludge sample

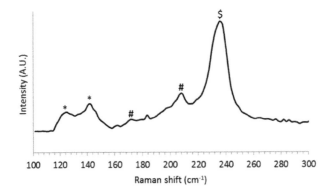

Fig. 9.3 Raman spectra of anaerobic granular sludge after the Se/Te-containing wastewater treatment. (*) show peaks assigned to pure metallic Te(0) at the expected positions at ~123 and 142 cm^{-1}. (\$) indicates the peak for trigonal-Se at ~ 235 cm^{-1}. Peak broadening and splitting into several narrow distinct peaks are visible. (#) peaks at ~206 and ~ 171 cm^{-1} probably indicate the formation of Se/Te alloy

9.3. Future perspectives

We have demonstrated the continuous synthesis of biogenic Te(0) and Se/Te alloys, but it is important that new studies emphasizing the optimization and scale up of these bioprocesses, particularly for CdSe and CdTe are carried out. The search for scalable and environmentally friendly chalcogenide NP synthesis processes addressing economic considerations (including cost-benefit analyses versus more conventional synthesis routes) and the impact of scale-up parameters on the structure and properties of the biogenic metal chalcogenide NPs are much required. Evidence showed that biosynthesis of chalcogenide (Ch) NPs could occur extracellularly as well as intracellularly. Certainly, additional studies are required to improve our understanding on the chalcogen reduction mechanisms, the location of the reduction sites and export mechanisms (in case of intracellular reduction) in microbial chalcogenide synthesis.

Biological synthesis of Ch NPs has several challenges, mainly regarding the control over size and shape, and scale-up of the process for bulk preparations, yet to be addressed prior to contemplating commercial scale applications. To improve the yield of synthesis and monodispersity of nanoparticles, factors such as microbial cultivation methods and downstream processing techniques need to be improved. The identification of specific genes and characterization of enzymes involved in the biosynthesis of nanoparticles are also required. Few studies reported to use AQDS as redox mediator to increase the reduction rate, but more research is required on electron-shunting pathways involved in the chalcogen conversion in microorganisms. Importantly, recent reports showing that with proper strain selection or with the inclusion of genetically engineered strains overexpressing specific reducing agents, the microbial synthesis of Ch NPs envisage as a promising field of research (Ramos-Ruiz et al., 2016).

Biogenic Ch NPs are associated with biogenic capping materials or EPS which enhance the compatibility and stability of the NPs under environmental conditions. EPS are specific to the microorganisms used for biosynthesis and we have shown the change in chemical composition or in fingerprint of EPS after Se/Te-containing wastewater treatment. But further research into the detailed characterization and optimization of the role of EPS in Ch NPs formation is warranted. The elucidation of the exact mechanism of secretion of proteins as capping agents due to stress conditions at the molecular level may eventually foster better control over size, shape, and crystallinity as well as monodispersity in the future.

The implementation of these nanoparticles (e.g. CdSe, CdTe, and PbSe) is, however, hindered by the presence of toxic elements, like Cd or Pb, as their release into the environment is inevitable. One of the main reasons for CdSe/CdTe QDs cytotoxicity is desorption of Cd (i.e. QD core degradation), free radical formation, and interaction with intracellular components or bioavailability (uptake) of QDs. In **Chapter 8**, it is demonstrated that biogenic nano-Se (nano-Se[b]) synthesized by anaerobic granular sludge was 10-fold less toxic than chemically synthesized nano-Se (nano-Se[c]). It indicates that the presence of EPS increases the physiochemical stability of nano-Se[b] and prevents their dissolution. The presence of EPS on the surface of nano-Se[b] plays a major role in lowering the bioavailability (uptake) and toxicity of nano-Se[b]. Still detailed studies are required on the toxicity of biogenic CdSe/CdTe NPs and the role of EPS on it. It is also important to focus on biological synthesis of "Cd-free" QDs (e.g. CuSe/CuTe) in the near future.

References

Bhatia, D., Bourven, I., Simon, S., Bordas, F., van Hullebusch, E.D., Rossano, S., Lens, P.N.L., Guibaud, G. 2013. Fluorescence detection to determine proteins and humic-like substances fingerprints of exopolymeric substances (EPS) from biological sludges performed by size exclusion chromatography (SEC). *Bioresour Technol.*, **131**, 159-165.

Bonificio, W.D., Clarke, D.R. 2014. Bacterial recovery and recycling of tellurium from tellurium-containing compounds by *Pseudoalteromonas* sp. EPR3. *J Appl Microbiol.*, **117**(5), 1293-1304.

Dzhagan, V., Valakh, M., Milekhin, A., Yeryukov, N., Zahn, D., Cassette, E., Pons, T., Dubertret, B. 2013. Raman- and IR-active phonons in CdSe/CdS core/shell nanocrystals in the presence of interface alloying and strain. *J Phys Chem C.*, **117**, 18225-18233.

Iovu, M., Kamitsos, E., Varsamis, C., Boolchand, P., Popescu, M. 2005. Raman spectra of As_xSe_{100-x} glasses doped with metals. *Chalcogenide Lett.*, **2**(3), 21-25.

Li, G., Cui, X., Tan, C., Lin, N. 2014. Solvothermal synthesis of polycrystalline tellurium nanoplates and their conversion into single crystalline nanorods. *RSC Adv.*, **4**, 954-958.

Li, S.W., Zhang, X., Sheng, G.P. 2016. Silver nanoparticles formation by extracellular polymeric substances (EPS) from electroactive bacter. *Environ Sci Pollut Res Int.*, **23**(9), 8627-8633.

Li, W.W., Yu, H.Q. 2014. Insight into the roles of microbial extracellular polymer substances in metal biosorption. *Bioresour Technol.*, **160**, 15-23.

Mal, J., Nancharaiah, Y.V., van Hullebusch, E.D., Lens, P.N.L. 2016a. Effect of heavy metal co-contaminants on selenite bioreduction by anaerobic granular sludge. *Bioresour Technol.*, **206**, 1-8.

Mal, J., Nancharaiah, Y.V., van Hullebusch, E.D., Lens, P.N.L. 2016b. Metal Chalcogenide quantum dots: biotechnological synthesis and applications. *RSC Adv.*, **6**, 41477-41495.

Raj, R., Dalei, K., Chakraborty, J., Das, S. 2016. Extracellular polymeric substances of a marine bacterium mediated synthesis of CdS nanoparticles for removal of cadmium from aqueous solution. *J Colloid Interface Sci.*, **462**, 166-175.

Ramos-Ruiz, A., Field, J.A., Wilkening, J.V., Sierra-Alvarez, R. 2016. Recovery of elemental tellurium nanoparticles by the reduction of tellurium oxyanions in a methanogenic microbial consortium. *Environ Sci Technol.*, **50**(3), 1492-1500.

Sadtler, B., Burgos, S.P., Batara, N.A., Beardslee, J.A., Atwater, H.A., Lewis, N.S. 2013. Phototropic growth control of nanoscale pattern formation in photoelectrodeposited Se-Te films. *Proc Natl Acad Sci USA*, **110**, 19707–19712.

Soda, S., Kashiwa, M., Kagami, T., Kuroda, M., Yamashita, M., Ike, M. 2011. Laboratory-scale bioreactors for soluble selenium removal from selenium refinery wastewater using anaerobic sludge. *Desalination*, **279**, 433–438.

Sridharan, K., Ollakkan, M.S., Philip, R., Park, T.J. 2013. Non-hydrothermal synthesis and optical limiting properties of one-dimensional Se/C, Te/C and Se-Te/C core-shell nanostructures. *Carbon N Y.*, **63**, 263–273.

Stylo, M., Alessi, D.S., Shao, P.P., Lezama-Pacheco, J.S., Bargar, J.R., Bernier-Latmani, R. 2013. Biogeochemical controls on the product of microbial U(VI) reduction. *Environ Sci Technol.*, **47**, 12351-12358.

Tang, G., Qian, Q., Wen, X., Chen, X., Liu, W., Sun, M., Yang, Z. 2015. Reactive molten core fabrication of glass-clad $Se_{0.8}Te_{0.2}$ semiconductor core optical fibers. *Opt Express.*, **23**(18), 23624-23633.

Tourney, J., Ngwenya, B.T. 2014. The role of bacterial extracellular polymeric substances in geomicrobiology. *Chem Geol.*, **386**, 115-132.

Tripathi, K., Bahishti, A.A., Majeed-Khan, M.A., Husain, M., Zulfequar, M. 2009. Optical properties of selenium-tellurium nanostructured thin film grown by thermal evaporation. *Phys B Condens Matter.*, **404**, 2134-2137.

Wang, Q., Kang, F., Gao, Y., Mao, X., Hu, X. 2016. Sequestration of nanoparticles by an EPS matrix reduces the particlespecific bactericidal activity. *Sci Rep.*, **9**, 21379-21389.

Wei, S., Cui, H., Jiang, Z., Liu, H., He, H., Fang, N. 2015. Biomineralization processes of calcite induced by bacteria isolated from marine sediments. *Braz J Microbiol.*, **46**(2), 455-464.

Xue, F., Bi, N., Liang, J., Han, H. 2012. A simple and efficient method for synthesizing Te Nanowires from CdTe nanoparticles with EDTA as shape controller under hydrothermal condition. *J Nanomater.*

Yuan, H.P., Zhang, J.H., Lu, Z.M., Min, H., Wu, C. 2009. Studies on biosorption equilibrium and kinetics of Cd^{2+} by *Streptomyces* sp. K33 and HL-12. *J Hazard Mater.*, **164**, 423-431.

Zhu, L., Zhou, J., Lv, M., Yu, H., Zhao, H., Xu, X. 2015. Specific component comparison of extracellular polymeric substances (EPS) in flocs and granular sludge using EEM and SDS-PAGE. *Chemosphere*, **126**, 26-32.

APPENDIX 1

SEM-EDX

The sample was prepared by spreading a small amount of nano-Se[b] suspension over a piece of a carbon coated paper. The sample was dried at room temperature and mounted on an aluminum holder for SEM analysis (JSM-6010LA, JEOL, Japan). EDXS analysis of the nano-Se particles was conducted using silicon drift detector (SDD, JEOL) technology attached to the SEM.

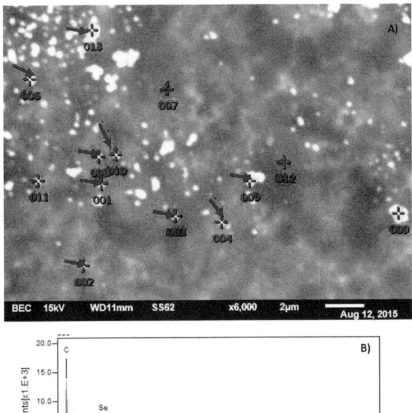

Fig. S8.1 SEM images of A) nano-Se[b], B) representative EDX spectra confirming the presence of selenium in nano-Se[b] (red arrow marked showed presence of selenium)

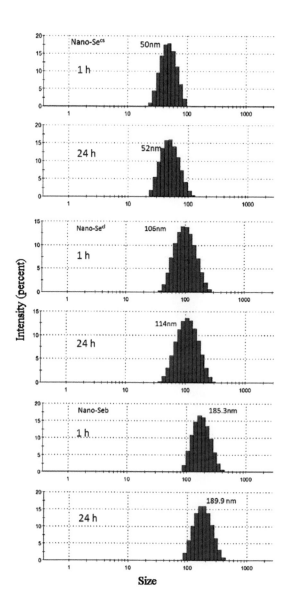

Fig. S8.2 Particle size distribution of nano-Sc^{cs}, nano-Se^{cl} and nano-Se^{b} measured by DLS after 1 h and 24 h of incubation in egg water (n=3)

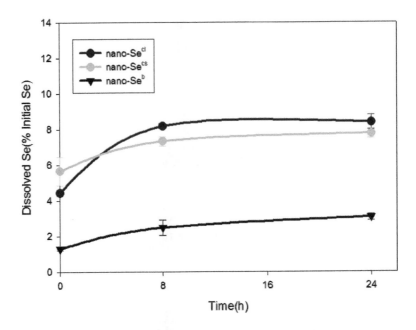

Fig. S8.3 Percentages of dissolved Se released from nano-Se^b, nano-Se^{cs} and nano-Se^{cl} suspensions prepared in egg water medium. Initial Se concentration was 1 mg/L (n=3)

Biography

Joyabrata Mal was born on 5th April, 1987 in Kolkata, India. Joy did his bachelor (B.Tech) in Biotechnology from West Bengal University of Technology. He later joined Indian Institute of Technology, Guwahati, India for his master (M.Tech) of Technology degree in Biotechnology. During his M.Tech time, he did his specialization in Environment Biotechnology and worked on synthesis of biohydrogen from carbon monooxide. Joy got admitted into Erasmus Mundus Joint Doctorate Program on Environment Technologies for Contaminated Solids, Soils and Sediments (ETeCoS[3]) and started as PhD Fellow at UNESCO-IHE from October 2013. As part of program, he also carried out his PhD research at Université Paris-Est. Joy also worked at University of Limoges under a short term scientific mission (STSM, COST Action ES1302). His research was mainly focused on developing novel methods for chalcogens (selenium and tellurium) bioremediation, bioreduction and recovery. He is currently a postdoctoral fellow in University of California Berkeley and working on selenium biogeochemistry.

Publications

Mal, J., Nancharaiah, Y.V., Bera, S., Mashewari, N., van Hullebusch, E.D., Lens, P.N.L. 2017. Biosynthesis of CdSe nanoparticles by anaerobic granular sludge. *Environmental Science: Nano*, 2017, 4, 824

Mal, J., Veneman, W., Nancharaiah, Y.V., van Hullebusch, E.D., Peijnenburg W., Vijver, M.G., Lens, P.N.L. 2017. A comparison of fate and toxicity of selenite, biogenically and chemically synthesized selenium nanoparticles to the Zebrafish (Danio rerio) embryogenesis, *Nanotoxicology*. 11, 1, 87

Mal, J., Nancharaiah, Y.V., Mashewari, N., van Hullebusch, E.D., Lens, P.N.L. 2017. Continuous removal and recovery of tellurium in an upflow anaerobic granular sludge bed reactor. *J Hazard Mater*. 327, 79

Mal, J. Nancharaiah, Y.V., van Hullebusch, E.D., Lens, P.N.L. 2017. Biological removal of selenate and ammonium by activated sludge in a sequencing batch reactor. *Bioresour Technol*. 229, 11

Mal, J., Nancharaiah, Y.V., van Hullebusch, E.D., Lens, P.N.L. 2016. Metal Chalcogenide quantum dots: biotechnological synthesis and applications. *RSC Adv*. 6, 41477

Mal, J., Nancharaiah, Y.V., van Hullebusch, E.D., Lens, P.N.L. 2016. Effect of heavy metal co contaminants on selenite bioreduction by anaerobic granular sludge. *Bioresour Technol*. 206, 1.

Pakshirajan, K., **Mal, J.** 2013. Biohydrogen production using native carbon monoxide converting anaerobic microbial consortium predominantly *Petrobacter* sp. *Int J Hydrogen Energy*. 38 (36), 16020.

Conferences

Mal, J., Bourven, I., Simon, S., Nancharaiah, Y.V., van Hullebusch, E.D., Guibaud, G., Lens, P.N.L. Effect of CdSe formation and accumulation on EPS fingerprint extracted from anaerobic granular sludge" at COST Action ES1302 - conference, 2016, Zagreb, Croatia.

Mal, J., Nancharaiah, Y.V., van Hullebusch, E.D., Lens, P.N.L. Microbial synthesis of selenite in presence of cadmium. Paper presented at Proceedings of the 4th International

Conference on Research Frontiers in Chalcogen Cycle Science & Technology (2015) in Delft, The Netherlands

Mal, J., Nancharaiah, Y.V., van Hullebusch, E.D., Lens, P.N.L. Effect of heavy metals on microbial selenium reduction. Paper presented at Selen 2014 conference at Karlsruhe Institute of Technology, Germany

Printed and bound by CPI Group (UK) Ltd, Croydon, CR0 4YY

22/10/2024

01777639-0002